T5-BQA-129

COME ON, NOW!
A PERSONAL JOURNEY THROUGH THE WORLD
OF THE AROMANIANS

by
Irina Nicolau

English Translation by Sorana Corneanu

EAST EUROPEAN MONOGRAPHS, BOULDER
DISTRIBUTED BY COLUMBIA UNIVERSITY PRESS, NEW YORK

2002

EAST EUROPEAN MONOGRAPHS, NO.DXCVII

Copyright © 2002 by Irina Nicolau
ISBN 0-88033-495-9
Library of Congress Control Number 2002101921

Printed in the United States of America

*I have written this out of love
for my father, Pavel Safarica*

FOREWORD

This book is published posthumously just a few weeks after the author's premature death. It marks the end of a "personal journey" not only through the world of the Aromanians but also of a world that "dissolved before they even left it." Yet, even if Phoenix does not look the same after it is born again, Aromanians and their culture have survived.

In a world tormented by ethnic, religious, and nationalist conflict, Irina Nicolau's recollections and reflections are pertinent to understanding the social and political dynamics of the Balkans. "Come On Now!" is a sensitive, often moving but ever relevant, tale by a distinguished folklorist and ethnologist who understood the past as well as the present.

Stephen Fischer-Galati

Have you heard that Greek saying? It goes, *"opios viazete scondafti"*, which means, *"whoever is in a hurry, this or that will befall him."* Since I have never been able to find out what exactly *scondafti* means, I imagined that anything can befall a hurried man. This is how I discovered how worthwhile it is not to rush into saying things, and to try and find the roundabout phrases that turn the way to an end into a pleasant stroll.

Stay with me, I need to feel you're by my side. It's been fifteen years since I began my book on the Aromanians. And now I find it hard to start work again; a whole lot of things have slipped out of my memory and there is only a handful of crumbs left of my former obsessions. I no longer see Aromanians wherever I turn my head and I am less prepared to turn upside down truths that have been established in a century and a half's worth of studies. Things were different then, I was kind of moonstruck. I was seeing things.

From time to time, I never knew when it was about to happen, I could feel a millenium of history surging up inside me. Or rather, I could feel an apple dropping through a split bag. Tens of millions of people, with their sheep, their gold, with their words and thoughts, sprouted up in front of my eyes, the Aromanians of all times became present and intelligible, and it was my duty to ask them questions. The experience was devastating. I felt responsible for each and every omission. I was trying to formulate all questions at the same time. Quick, right now... I couldn't. At the last call, I would decide to ask one person one single question. But before I could choose the most appropriate words, and before I could decide who the interlocutor should be, the image would painfully fade away. I would immediately start to put things down on paper. I was dreaming of writing a book with pockets and whiskers... I was experiencing unusual emotions, my soul was aching.

Still, when I look into the past, I can see that the origin of my aching goes a long way back. I was four when my father used to take me to Bardaca's Coffee-house, behind the National Bank. There we passed among the tables, father holding my hand, and he told everyone, "Look, this is my daughter, Miss Aromania." There were only Aromanians there, and they knew father had married a Greek woman. In order to test my feelings, they inevitably asked me one question, *"Tze hi tini?"* That is to say, "what are you?" And I would reply, in Greek, that I was Greek. A horrible thing to say! They were all infuriated, and father said it was all well.

Much later, when I was coming of age, I sensed how lonely father felt in a house he shared with three Greek women, and I decided to move on his side. Unawares, I would find myself repeating in my head: "I am an Aromanian, I am an Aromanian, for my father's sake I am an Aromanian." The fit of solidarity went as quickly as it had come. But it so happened that once, in winter time, one morning at eight o'clock, St. Nicolae filled our shoes with tears and funeral candies. Father was dead. Pavel Safarica left behind a whole world. I was not sure what I was going to do with it. When it comes to heritages, this is how things are: you either accept them or not. If you do, you answer for what becomes of them, if you don't, that is betrayal. The thing that solved the dilemma for me was a research topic; thanks to it I discovered that my father's world was mine, too.

Every time I start writing about the Aromanians, I feel my soul inhabited by the wintering presence of my father. Come, father, sit down! I've brought the basin, let me wash your feet. Here's some meat and there's some fresh water for you over there. All is well again... I have started to write your book. Couldn't you stay, till the end of life and till the end of death?

Dear reader, it seems only fair that I should tell you what to expect. Be ready to encounter a world of broken images. Try to imagine that I was once a diver and that I discovered Atlantis on a seabed whence I returned and forgot what I had seen. All I can be sure of is the dirt from under my nails. I have scuttled the floor of Atlantis. Nothing more! Or imagine that I am Semiramis and that you are my guest for a couple of hours. You expect me to open the superimposed gardens for you, and you only get some hotchpotch. Eat! We no longer grow flowers – only vegetables. Yet there is a garden for you to discover in every bite. Have some more hotchpotch, if you love me, it tastes good...

The Greek women in my family never put on any make-up, and they took pride in their homeliness. One of my aunts, though, knew how to use lipstick, and would paint herself wherever and whenever she had the chance to. The lack of a mirror was no impediment for aunt Pepo. Her lipstick in her right hand, she would look at her face in the left palm and draw, with dream-like accuracy, two horizontal brackets of cyclamen hue. I was a child then and would peep at the "mirror" in the bare palm, but could only see the short line of her life. This was all too disturbing. "Why don't you go all the way," I'd cry at her, "let's try it without the palm, come on!" To make her do it, I'd hold her left hand and press it against my bosom. "I can't," my aunt would protest. "Yes, you can!" And then the lipstick trespassed the old frontier of her lips and drew, in thick greasy paint, the funny mouth of a clown.

I believed for a while that the rectangular mirror that she rarely took out of her purse had transferred part of its mirroring power to her. Then, at some point, I imagined that her palm was the prop point of her eyes and that, while actually looking nowhere, she drew a mental contour which her memory had once stolen from the mirror.

You might wonder what Pepo is doing in a book on the Aromanians. I'll explain: she simply sprang out of me. While she lived, she loved being watched. I told myself: she has no children to remember her, so let her stay. I could have easily found a legitimate place for her in this book, so that no one can complain: what's this Greek woman doing here? After all, she was married to an Aromanian. Nacu, her man, was as tall and handsome as an Albanian, and he loved to sing as the *farseroti* do. At celebration time, when the whole family was gathered, he'd tell Gheorghe and myself: "You just keep the tune like this: iiiiiiiiii..." And with our accompaniment, he started singing:

"*Hai, moi, ş-d-iu ştia a greţli*
S-bea, moi, ap-araţi?"

That is to say, how come the Greeks know how to drink cold water? In the next lines the Greek knew nothing about drinking wine or eating bread and meat, they had learnt it all from us, the Aromanians. Mother would turn moderately angry and invoke the ancient Greeks' wine trade while Pepo cried bitterly.

I promised I would take you along a road where you would see Aromanians, but don't forget, not for one moment, that we are out for a stroll. I've deliberately chosen no precise role for Pepo. I simply want her present, in no meaningful way. See? There is this crazy notion taking hold of me that my chances of writing about the Aromanians are the same – wherever I start from and no matter what the ending may be.

Something happened in the winter of 1831 that's worth noting, not so much for its own sake, as for the lesson to be learned from the conduct of a man worthy of the name.

Alexis Barda's manner was that of a Turkish army leader. His word was like the ram's bell among the sheep. That autumn, his mind was set: "we're wintering in Casandra!" And no one dared to cut him short. On the road, obedience had to be mute and blind, democracy is for the sedentary.

Twice a year, the storks and the Aromanians provoked shouts of joy and hope in Casandra. The Greeks' white houses absorbed the mutton smoke and their dry wood chests gulped down the silver coins of the tenants. *Calimera, calispera* and communication went further than that, though. When people come together, souls do, too. The days passed. The women, children and old people lived in the village. The men were with the sheep or away, attending business. One night, the wood pieces and pots propped against the doors were scattered like feathers all over the place. Caramiciu's thieves overran the houses rented by the Aromanians. In a twinkling of an eye they pilfered all the gold and silver pieces and took six married women with them, leaving word that they expected a ransom of 80,000 *grosi*.

Exactly how old Barda learnt about it all is of no consequence. He gathered several men and rode off to find the thieves. I can imagine him dismounting near Caramiciu's cave, grinning like a samurai. *"You, conniving ape,"* he would roar in imperial anger, *"you had no other business but to steal my things and break into six homes. I gave you sheep when your men were hungry, I gave you money for your clothes and guns, and you, foul snake, what do you do? Steal from me! Shame on that father of yours that gave you life!"*

With a swift, unprepared gesture, he then slapped him twice, and twice again. Then, he emptied a purse full of golden coins at his feet. *"If you wanted money, why didn't you come and ask me for it?"*

Considering things settled, he looked at the six women who were standing aside, eyes fixed on the ground. *"Come!"* The gold and silver, together with the women, took the way back to Casandra. Later on, Alexis Barda died, poisoned by someone who did not dare to look him in the eyes when he killed him. It was the time when our people were beginning to give up their mounted life.

I have been personally acquainted with some of Alexis Barda's descendents, at a time when they were already called Badralexi. At a first glance, one could easily mistake them for gentlemen. I realize now they were not. Under their well-cut clothes their bodies were too lively, their voices too powerful. In thinking about them I understood that the quality of gentleman is the reward of he who returns after a long journey, a journey that refines the soul and slenders the body. Yet, when I'm saying they were not gentlemen I mean they were something else, of equal value.

Our godfather, for instance, loved to put on the exquisite gait of a thoroughbred horse. Many thought he was lording it. Actually, what I've noticed in all Badralexis was a certain natural reserve, slow motions, something of the easeful manner of big animals, which makes, by comparison, the fretfulness of small animals look all the more ridiculous. When godfather Paul first met my mother, in Galati, at her wedding, he told her: *"Don't be angry with us, Angela, we are the first generation to wear shoes..."* And he added, somewhat amused: *"after all, you are the second."*

If wearing shoes means shuffling them through school halls, then godfather Paul was right. Yet, if one counts the *celnic* shoes as well, it turns out that the Badralexis had long been putting them on. When Alexis Barda's daughters, Sirma and Tasa, were to be married, he looked for bridegrooms up to Avdela and Baiasa, and he would have reached the Equator, too, if that was what it took to find men who would measure up to himself.

That is why I cannot really understand why Caramiciu the thief did what he did, when he tried to plunder Casandra. I'm trying to picture him: short, rash, foolish even, the lines of his face a bit asymmetrical, knit eyebrows and a cunning look... How was it possible that he'd be slapped in the face at his own "place", in front of their thieves' cave. It may be that the *celnic*'s resoluteness and calm, the one moment's devastating fury and the way he resumed his poised role of master, were weapons the other could not handle. Then it's even more confusing – what in the world could bring him to lay hands on what belonged to Barda by right? Barda, from whom he had received food and money? Right now I can think of nothing else but that he simply acted foolishly, that the thief made a mistake.

It is all even more confusing as the lists of Aromanian captains drawn by Ioan Caragiani mention the name Caramiciu twice. Who knows whether our man was not the son of one of them! The theft the story tells of may have been the only one in his whole life, he may have been in love with one of the women he stole, or one of Barda's men may have offended his father. Hard to tell. Anyhow, I declare myself on Barda's side, who, at the beginning of the previous century, actually after 1831, settled the foundation for Badralexi's Calive, a hamlet not far from the village where my father was born.

Coming back to Caramiciu, I will say that an idiotic petty thief cannot mar the image of the Aromanian mountain fighter. At some point, the terms *"armatol", "cleft", "palicar",* and *"comitagiu"* were synonymous and denoted the rebellious men of war. How they came to gather in the mountains, that's another good story to tell. Pouqueville says that they did so in accordance with some age-old military structures called "captain-shires". The Turks, who could only be pleased if order was preserved in the region, did not destroy them. Leadership of the captain-shires was hereditary. Caragiani notes that it could also be passed on by way of dowry. There were, at some moment, up to 180 captain-shires. The sultan acknowledged them, and the Turks addressed the captains as "bey" and "pasha". When the Turks started to replace the Aromanian captains with Greeks and Albanians, some of our men, outraged, withdrew to the mountains. The *celnici* supported them with food and money. They gave them little lest they should go elsewhere and take more. The *clefti's* calendar and the calendar of the Aromanian cattle-breeders counted two seasons: the summer-time, from St. Gheorghe to St. Dumitru, and the winter-time, from St. Dumitru all the way back. Winter on the plain, summer in the mountains. I owe Fauriel a detailed description of these controversial characters. He makes no distinction between the captains and the *clefti* in his writings. Fauriel maintains that someone could give up his status as captain for the position of a *cleft* and the other way round, within the compass of one year, for, he writes, those who opted for the resistance remained bounded to the *armatolic* they came from. As the Turks were willing to protect the territory against more damaging thieves, they tried to win the *armatoli* over to their side, by offering them guardianship of the mountain passes. The position was well-paid. This is how a distinction was created between the savage *armatoli* and the tamed *armatoli*. The savage sort could be told from the others thanks to a red woolen belt they used to immobilize their prisoners. As for the rest, Fauriel shows that either sort wore a gun, a sword, a knife and an armor piece that covered their chest, and silver sheaths tied to their knees with leather belts. The Frenchman deplored the lack of any military technique and claimed that even the most famous fighters fired their guns and pistols at random, from a standing, kneeling or lying position; that they hid behind trees and corpses, just like children would; that they favored, above all, the surprise element, and got out of a blockade by hand-to-hand combat. He also admiringly observes, though, that Nico Ceara would jump over seven horses in a row, that some of them were able to jump over three wagons full of thorns and that others could run faster than horses. He also says that all of them could easily endure hunger, thirst, pain and fatigue, that they would rather die than be caught and that they knew of no bigger shame than seeing their captain captured.

For those who are not familiar with it, Fauriel's book is an anthology of songs. The Albanians called these songs *"bucovale"*, from Bucovala, an Aromanian *armatol* who, after a fair bit of fighting in the Balkans, joined the Russian military and died – no one knows how and why – in Jerusalem.

Captain Spiru, my Greek grandfather on my mother's side, passed on to me the following story. It is said that the Aromanian *armatoli*, who had been the heroes of the fights in Misolonghi, gathered one day to decide which of them was to leave for England. Lord Byron had spoken in praise of them (he believed they were Greek), and the King or Queen, whatever they had at that time, had invited them over to London. One said this, another that, until, finally, their captain, having heard enough, said: "Enough with this blabber, England this and England that! You'd better have a look at yourselves..." and he fixed a critical eye on their uncut nails, their crumpled cheeks, expensive yet dirty clothes, the sandals full of mud. "Look at yourselves," he repeated, "and keep your place. It's better people hear of us, we've nothing to show."

That is why I believe that unholy deeds like Caramiciu's cannot shake off the image of those *palicari* with moustaches and shining clothes woven in silver and gold, who stand still in the portraits of the History Museum in Athens. They might well have been thieves, but they made their entrance into history through the front door, carrying with them a ragged patch of land from which the new Hellad was made. But it wasn't for the Hellenic Kingdom that my Aromanians had fought. Aged and embittered, they were to explain, later, over a cup of something at a coffee-house table: *"We meant to set up a rumeic, which is to say a multi-national state on the model of Byzantium. This is what we wanted and we ruined our people instead!"*

Look for them and you will find them everywhere! I met Aromanian *armatoli* when and where I least expected to. As a result, I had to give up the patriarchal image of the warrior smelling of mutton. They often reached the most distant places, flying on vultures' wings. Here's one example: the Chamberlain of Empress Ecaterina II of Russia was an Aromanian from Seatistea called George Papazoglu. And his word weighed a lot too, since he persuaded the Empress to send Admiral Orlov to Greece in order to urge the population against the Turks. It all happened in 1765, i.e. 56 years before the Eteria. Or, if Papazoglu had anticipated the Eteria by half a century, and Rigas – the Aromanian from Velestin – drew its manifesto, if, after all, the Aromanian *armatoli* fought with arms and weapons for the liberation of Hellad and they won, do you realize what a mighty slice of history's melon should have been ours by right? Well, I'm telling you, we got nothing. Not even the seeds!

William Martin Leacke had the chance (how I wish it could have been me!) to be present, as he was himself the guest of a Vizier, at the welcoming of several Aromanian *celnici* who'd come to bring the tribute, more exactly the taxes for the sheep, just like in Solomon's time, for the right to make timber, to fish etc. Which was actually immaterial to the case of Sultana Validé, the woman in the harem who had the honor to be the Sultan's mother.

The money had to be there on time, the *celnici* left their mountain abodes for that particular purpose. Now, this is what I believe the Englishman saw, for this is what it's all about. At the Vizier's court, in rooms where the air was full of fragrances and the walls covered by carpets, the visitors walked on tip-toes, bowing and curtseying, kissing some coat tail or ring, slipping in a word or casting a glance. When the Aromanians came, the man in charge with announcing the new-comers did so. At the thought of those brutal, fierce-looking men who smelled of mutton, the Vizier would have said: *"Don't let them all come in, just bring me their leader!"* *"Let only the leader come in!"* the Turk passed on the word, but after a few moments, he came back and said: *"I told them what you said and they answered –* the Englishman notes *– 'We are all equal'."*

What a strange sort of people we were, that's what I say! The leader decided for everyone else, the old man was above the young man, yet, all of a sudden, *we are all equal*, how can I make sense of this? Another thing to notice is that the Aromanian dialect lacks the pronouns of politeness and that, at some point around 1920, any old shepherd dared speak quite freely to teacher Gheorghe Celea who, in Gramaticova and its whereabouts, was considered a sort of king. Celea's intelligence, education, and commitment to the community had turned him into a "king", yet the shepherd's old age made of him a consecrated emperor.

The *comitagii* appeared later in history, playing the part of fallen angels. Nevertheless, their career sometimes inspired the others with respect: for instance, Sultan so-and-so granted a pardon to Caiu the bandit, who had pilfered Mount Tomor some twenty years before, on account of the impressive fact that he had never been caught.

In 1879, band head Vasile Jurcu, together with 25 *farseroti*, tortured six people and stole 100 *lire* from them. The next day, near Comati village, two bandits came out and asked for 50 *bani* per cattle head. The shepherds got angry and killed them.

Another story. On February 4, 1905, around two hundred men set fire to Hagigogu's sheepfolds near Caterina. The officer who led the two hundred bandits was called by a flower-like name, Amalia.

But there was also a sort of code of honor that the bandits respected. They did not kill people unless there was no other way. Usually, they fired at the horses' feet. When thrown off the horse, the man was no longer to be feared.

Burileanu records another story, wherefrom we learn what it was like to die in the Balkans. The man who is about to die was an Aromanian, and he was attacked by four Moslem Albanians. He took one down, but then he himself was brought to the ground by another's bullet. When the brigands came nearer to him, thinking he was dead, our man took this chance and fired his gun at one of the remaining three. The others got angry and shoved two bullets into him, one each, and then went closer again, but he wasn't dead, so he seized his gun with both hands and whacked one of them in the head. It was the end of that man. Obviously, the surviving thief shot him dead and probably remained to tell the story, too.

Fauriel and others claim that in the Balkans, guns rarely reached their target. They also say, though, that in order to test their ability, the men tossed a coin up high and tried to shoot it in the air. I cannot but conclude that some were good shots, and some others bad. One fine shot was, for instance, Mitru Vlahu, whose photograph I could see in Burileanu's book. A short man, with long hair and a black beard, a Robinson with *hamaili* – that is, small silver icons hung around the neck – he was a damn' good shot. He was once in a house in Kastoria, under siege, and after he shot a Turk, who had come too close, right in the forehead, he saw another Turk, who was their commander, raising his trumpet to his mouth in order to signal another attack, and he aimed so well that the bullet went through the trumpet and hit the man right in his mouth. He did that alright!

They say that Sipisca sheltered a nest of bandits who reigned over the whole village. On St. Gheorghe's Day they gathered together, just like for an Olympic event, and roasted a living ram.

Alice, our goddaughter, was born into the Cocones family. She has no idea what a fierce ancestor she's left behind. A "Greek-maniac" – which is to say an adversary to the tendency of befriending Romania

– he was a brigand with a moustache as dark as a roasted chop. His protection was valued as highly as that of the Sultan. Here's a dialogue recorded in Burileanu's book: *"See this gun? No one here is good enough to have it, it's a soldier's gun. It was I who laid hands on it, and it is my business how I did it, and I can very well choose to walk up and down* Pogradet *with it on my shoulder, and no man is to ask why. This is the manner of those who bear the Cocones name. We don't even pay the tax. For no one would dare ask us to pay it. If anyone should harm me, they're sure to get a bullet from my brother and were he harmed as well, another brother of ours will come and settle things. There's many of us, and we all grew up with a gun by our side."* The style of the book avoids such a grand manner throughout, which makes me believe that the author faithfully reproduced what the brigand told him. He listened to him, then explained how unwise it was to be a Greek-maniac, for, don't you see, there is your Romanian people whom you have to defend. On hearing this, Cocones reckoned – and here I quote again: *"Hold that tatter, man! What can I say? How should I know? What shall I believe? If you're brothers of ours and Romania is our mother, where have you been so far, why have you left us at the hands of the Turks and Greeks? We, the Aromanians, are brave people, but we are orphans too, we know of no father. If only our father lived, that would be a day!"* Before parting, Cocones kissed him on the mouth, as he would a brother, and "Miss" Burileanu, for I can't call him otherwise, washed his lips with water from the first spring he found, and rubbed them hard with his handkerchief, to wipe away the trace a killer's mouth had left on his.

One night I dreamt that I was sleeping and dreaming. In my dream I was on a green meadow. By my side, lying, was my body, which for some reason I couldn't see. And my body talked to me and said, let us fall asleep and let us dream together the same dream! But we cannot, I answered with regret, look, when I will dream of a hill – and above us a hill was drawn on the sky – you will dream of a valley. And it cannot be otherwise, ever. My Aromanians and your Aromanians, those I speak about and those you imagine, might not be the same. I've told you my dream to warn you.

In order to be a *celnic*, one had to have a *falcare*: people, sheep, goats, horses, mules. If too many people came to live in one place, things tended to go awry. When that happened, some of them would "hit the road". In the 11ᵗʰ century, Kekaumenos recorded the word *celnic* and translated it as *stratygos*, which means the man who leads soldiers in a fight. The etymology of the word is Slavic, *celo* means *fore*, but the *celnic* was more than one who is "to the fore". It was also close to *scuteris* (*ecuyer*), and it was a title pointing to military authority. Among the people of Gramoste, the *celnic* was called *chihaia*. Until the end of the 19ᵗʰ century, there were still real *celnici* around. As such, one was expected to be many things for the community: a leader, a judge, a healer, a priest, a translator... When someone needed to know how important one particular *celnic* was, he would ask, *"Is his* falcare *big?"* It's as if today you asked whether such-and-such has many subordinates.

There is this story that has come down in the Celea family. Lala Yioryi, otherwise known as Papu, was once approached by some men who did pro-German propaganda, so he could hear how Hitler was this and that and many other things. And Papu, in order to size up the whole story by another measure of things, asked them: *"OK, now you tell me, how many sheep does this Hitler own?"*

At some point I drew the family tree of the Celeas. It took Gina, Nicolae and me a whole year to do it and we came out with something like a full-sized bed sheet. Eight generations, hundreds of people spread over innumerable countries on three continents... To each name was appended a whole bunch of stories: whom he or she married, where they traveled, things they went through. At the point of origin – the furthest back memory could reach, the root of roots – was a founding figure, a woman. Her first name is lost now, they used to call her *Tal Celea*, which means *"she who is Celea's"*, a wife's denomination. Only a fragment of a story has come down to us.

After her husband's death she was the first who could gather the strength to say "Come on, now!" She was as stubborn as an old ram and as proud as an empress. Those who couldn't put up with a woman's leadership, be her in the family, left. The others learnt to follow a skirt: in summer, up in the mountains, in winter, on the plains near the sea. "Come and see the *celnic woman's falcare*," shouted the people in the villages they passed through. The handsome people in new clothes and their big herds of animals, were a thrilling sight. And she walked with a stately gait, ahead of the rest, in her red sandals with an upward pointed tip. Years on end, she kept repeating the words and gestures of her dead husband, at times even placing such weight on them as could only come from the madness of a female lion after childbirth, and so she raised her *falcare* above any other. One year, she sent no less than two hundred horses and as many mules to Ali Pasha of Ianina. A queenly gift which she was able to present, and the Pasha understood her message: "I can make such gifts to you, as you can well see, and in exchange I expect you to leave my people alone."

The *celnic woman*'s people were exempt from paying taxes. When their steps took them over the Vardar River, no bridge tax was asked for the cattle, the people and all their belongings. She alone had a say in who was to go first and who next; she stood there and ordered, now you go, now you, and you... And the heads of the families acknowledged and said: *"amin, moi, Tal Celea."* Which is to say: "so be it, as you say."

Tal Celea closed behind her the gate of a golden age. Hard times followed for her people. Her sons were slaughtered on Ali Pasha's death. Only one of them made his escape, a narrow one. Poor and alone, mounted on a lame mule, he became rich and renowned. People called him Celea Perifimos, that is Celea the Proud One. The story goes that he watered his horse on wine. Now, two hundred years after, the formula of acknowledgement uttered when crossing the bridge over the Vartar has become for the *celnic woman*'s descendents a mere joke, something like "OK, OK, as you please."

"There goes another day and we tricked it again..." This is what my grandmother used to say after dinner, while she cleaned the table. "It" was the ever-demanding belly. Man, though, was not to comply with it. Greed was a sore sin. They ate with their hands. Grandmother used four fingers, her two forefingers and her two thumbs. When she was done eating, she wiped them on a handkerchief. While eating, she was as neat as a cat, and would leave no crumbs. We sat at the table on the ground, with legs crossed under us, like Turks. The Aromanians in the villages had a sort of low tables, like those of the *rudari*. The nomads carried with them rugs of a special make which were called *masale*. The tablecloth is not a sign of vanity; man is different from the animals in that he eats on clean surfaces. The townspeople ate like the townspeople in the Balkans. We were famous as pastry-cooks. The Aromanians cooked a lot of food for other people.

In 1880, Chirita Rosca willed 100 *florii* to the church in Bitolia, 100 *florii* to the Macedo-Romanian Committee, 100 *florii* to the Girls' School for a fire pump and another 100 *florii* in solid gold coins to the Charity Trust, which was to provide dowry for a poor girl every six years.

In 1904, Ciumaga subsidized the building of a retirement home of 60-80 beds. Zega put up the money for the church and hospital, Badralexi for the drug store, Celea for the school and so on and so forth. Hundreds and thousands of *florii* were offered for roads, bridges, water-pumps, for the poor... Everybody put up some money; Sina did, Dumba did, Tosita did, they all did, even when they were abroad, in Vienna, Budapest, Odessa, or in Bucharest. This is the reason why the word *lasa* [Rom. for "to put up the money for"] came to mean, at some point in the history of the Greek language, *testament.*

Take my hand and let me show you what we did for Budapest and Athens. Look what buildings Zappa and Tosita and Averoff left behind. They are enormous. And you haven't seen Sina's and Dumba's. All were funded with money given, as father said, *jancoté* (on the spot). We were a very liberal sort of people, but nobody knows it any more. Giving was, for the Aromanians, a kind of *potlach*. In its original meaning, *potlach* refers to a h2abit of the primitive peoples – a way of gaining prestige by destroying goods. Let's say we are in Alaska and I say this to you: "I own 100 good sleigh dogs, I will kill 80 of them right now." And I start ripping the dogs one by one. You cannot pretend this is naught to you, you have to accept the challenge. You kill as many dogs as I did and some ten more if you want to be the winner. It all happens under the awed eyes of those who own no or few dogs, while we, standing on top of the pile of dog corpses, ruined in a couple of hours, exult in the prestige we've just gained. For he alone that owns things can afford the luxury of losing them, and luxury is a sign of power, see?

Now let us imagine that we are Aromanians and that you are Zappa and I am Averoff. We own many sheep, and most of all, many golden pieces. As for eating, we still enjoy a good loaf of bread with a cheese topping. Our bread is probably of a whiter sort, but how white can bread get? We don't cover our wives and daughters in gold, it would be shameful to do so. We are too noble to raise palaces for ourselves, so what can we do to have people talk about us? Here's what I say – a library in Athens! And you, you settle for a park!

And then I have another library raised in Mezzovo, for that's where I was born, and you think of something else, too, and so on... The splendor of our deed lies in that we are never going to see the works we've funded. The real Averoff refused to go to Athens and see the wonders that were built with his money, and Zappa covered the costs of the park in Athens, which bears his name, by clause of will, after his death. Others have enjoyed our gifts, all we did was dare the limits of generosity.

There, on the right side. The one with long shirt hems. See what a big moustache he has? His name was Constantin Mergeani. A grand figure of a man he was! Rich and respectable too! His people all but worshipped him. He had his weak point, though, and that was vanity. Too much vanity. See for yourself. We were in Silistra. One night, the forest guard killed one of his dogs. A good dog – that's something you need in a sheep herd. What can I say, baruti, *the old man worked himself up into a frenzy. Yet not because of the loss – it was his vanity. He hired a lawyer, the best one money could get, and God knows how, he managed to win the lawsuit; the lawyer's argument was that since the dog died, the herd could have been lost and then who was to be held responsible for that? The woodman! The forester had to pay for the herd. The judge translated that for him: "you owe Mergeani 20,000 lei."*

There was crying and there was wailing. I can make this story longer, but there is no need. The forester took out of his pocket five one thousand lei bills and said that was all he had and that he couldn't give what he had not. Grandfather stood there and watched the show of the other's shame and then took the money from his hand, turned towards the people in the room and said: "I don't need all that money from you. Take it back!" And counted four thousand and returned it. "As for this thousand here," *and he waved the bill above his head,* "I will buy food and drinks for everyone. And you," *he howled at the woodman,* "the next time you see the shit of my sheep, you shove it down your throat!"

"*That's what grandfather was like, an arrogant man*" – and Dinu Mergeani brought the story to an end. He had just turned 80 back then, in January 1986. He told it in the summer of that year, in Tulcea, in the yard of his house, while his wife, Sultana, was preparing an eggplant salad.

It was Mergeani, too, who told me this one, actually a simple anecdote: "*Several years ago I was in Bulgaria. I traveled all over the place. It's beautiful! Out of curiosity, I went to see a village we used to cross in winter time. That's where our people used to leave the things they couldn't carry all the way: good clothes, covers, even money. And while I was there, a man I knew, a Bulgarian, asked me:* 'how is Iancu, what is he going to do with the things he left me? Twice a year I take them out into the fresh air and spread tobacco over them to keep the moths away, they are good clothes, it'd be a shame to waste them.' *Iancu had trusted him with those things forty years before, and God knows what had happened and he couldn't take them back, and what that man meant was neither that he had had enough, nor that he was too old and could no longer take care of all that stuff, that there was no more room for it or anything, he simply was curious to know what Iancu was planning to do with those things he'd left with him!*"

The saying that goes, "*All there is in the chest we wear*" was for me the measure of a sparse living. An even sparser way of saying things would be: no chest at all. Although our *harari* – our woolen bags – could be taken for a sort of chest. You'd fill them with whatever you had and mount them on the donkeys. As objects became more numerous, people learned to scatter them around the villages they went through. They'd leave them with trusted friends, and they remained, from a distance, the owners of those things. They'd also leave money behind. "*Weren't you afraid at all?*" I asked Mergeani. "*Afraid of leaving it with friends? How could we? It was safer than a bank. A bank goes bankrupt, a man keeps his word.*"

Among us it was a shameful thing:
  not to kiss the priest's hand,
  not to be respectful enough towards your parents,
  to call your husband by his first name while there
    were other people around,
to do dishonest business,
to be seen by the fiancée's relatives before the wedding,
to be heard uttering foul words,
to disobey the aged,
to break your word,
it was a shame to marry someone from another kin-line,
or to receive with your dowry a mirror that was too large
    or too small,
to beat up your wife,
to neglect your children,
not to obey your husband's will.
It was shameful to be seen naked,
to beg, to lie,
not to work hard enough,
to eat food that was too tasty,
to let your enemy cut off your moustache,
to allow that your captain be caught,
to see how your horse's tail is cut off.
It was a shame to lie,
to quarrel with your neighbors,
to commit the carnal sin,
to prove incontinent
when you drank, when you played cards or when you dressed…
But above all it was a shame to have no shame.

    When, at the time of the communists, father used to take all kinds of jobs, mother asked: *"Pavel, have you no shame?"* *"I'd be ashamed to have an empty freezer,"* was his answer. Times were different then.

    Since I haven't seen it, I cannot swear it ever existed, and so much the less can I indicate the whereabouts of the cobbled street made by the drunkard subjects of the Sultan. Some say that it was in Bitolia, on the other side of the barracks, a winding line among the buildings of the Turkish neighborhood. Others claim it was in other areas of the town. Anyhow, it seems that Allah did not love the drunkards. Neither did the Sultan. That is why the drunkards were chased down to the tiniest nook and brought in front of the Cadi. *"Ti, ti, ti,"* the Cadi said, pulling at his beard, *" this man has broken the word of the Koran, take him down to do the cobbled stone!"*

It was a dire sentence, the same for all. Thrown out into the streets, beys and vagrants alike carved the big stones, thus turning a few hours' disorder into the lasting order of stone.

In Istanbul, the Turkish drunkard fared even worse. The first time he was caught dead drunk, he was clubbed to the ground and entered into a register. If he was caught dead drunk a second time, the clubbing was fiercer, blood was shed and the man was bedridden for several weeks. At any rate, he was entered into the register once again. Well, if the man was perseverant enough to get drunk a third time, it was the turn of Turkish tolerance to manifest itself, and the man was declared an *Imperial Drunkard.* As such, when he was found lying on the ground, he was handled with much care and taken to the nearest *hamam* where he was left to sleep in a warm bed of ashes. In the morning, the servants washed his clothes, he was cleaned and shaven, he received a free cup of black coffee and was returned to the street safe and sound, with a right granted by the Sublime Porte to start drinking all over again whenever he pleased.

In 1984, I went to the Annual Ball of the Aromanian students, which they organized at the Parc Hotel. I could see over two hundred boys and girls there who knew how proper drinking is done, which is without becoming intoxicated. They had the beauty of unspoiled souls. At that moment, I wished I had a country to trust them with.

Since so far I've been only talking of honor, community, generosity, you might think that we were a society of angels, unable to kindle tension, conflicts and hatred. It would be a terrible thing. A writer who knew us in the mountains in the 19th century testified that the quarrels among the *celnici* made the valleys resound. No one would interfere, not even the Turks. They were left to their own brawls, all by themselves. I came across several accounts on Paligora's kin where it is said that they were scattered around the world because of a number of fights they had. They left and were lucky, they got to earn money and social status. They were known to be a bellicose, angry sort of men. And so people say of such a person, "this is one of Paligora's, with the axe at his waist belt."

When the rupture affected people who were unusually close, which is to say when the break could not be accepted, a woman took her baby into her arms and went to the enemy, who was bound to accept the herald. Thus, a woman and an infant could reestablish the union that had been upset by the men. Among the Serbs, the meaning of the gesture was even more explicit, as the woman carried in her arms a child who had not yet received baptism, and cried from afar to the adverse family: *"Baptism, baptism!"* No one could be so wicked at heart as to refuse the Christian blessing to a newborn. Once they became related, the enemies went along with each other, for a while at least.

It should also be said that there were certain Aromanian kin-lines who knew vendetta; that among the members of the same kin there were relatives who would not speak to each other for tens of years on end; that there was no forgiveness for certain mistakes; that the adulterous woman was stoned to death, and the prodigal son chased away. Still, if one is venturous enough to look for them, one can find Aromanians among the thieves and the burglars, too. Whoever refuses to do that will miss the chance to utter a solid piece of truth, namely that in a closed society like ours, there were fewer rascals than anywhere else, and the community treated them with the hatred and disdain one feels for one's enemies.

If you were to ask me what it is that I most despise in the way the ethnic groups from the Balkans dress, I will hasten to answer: excess. For instance, can you imagine what that young Bulgarian woman looked like, whom Nenitescu describes as wearing around her waist a belt 25-metres long? In order to do it up, she went out in the street and someone helped her; and above the belt she put on a skirt, then another skirt, then the apron, the buckles, one, two, three pairs of them, and when she could no longer move and looked like a tub, her sweating face lightened up for she knew now she was beautiful. What about the Albanian guy who sewed his shirt with 300 gussets? What could such a shirt weigh, how much could it be, how was it washed? On top of it he put on the vests with gold and silver braids and laces, the belt where he hung the guns inlaid with turquoise – and when he got to look much like a tree in blossom, the Albanian, still ambitious to make himself even more handsome, touched up his walking manner, and made an artificial, wobbly, broken-hip appearance. Excess was also manifest in the finesse of the Greek lace, the spider-web-like frills and fairy-tale gazes the women wore in the Turkish harems.

One morning, I was in the Benaki Museum in Athens and my companion was a Greek man who was a perfect stranger to folk art. The only reason why he was there was to guide my steps, and at a certain moment, seized with wonder and pride, he asked me whether we, the Aromanians, had anything like that. His question hid no evil intentions, yet I gave a grim, furious answer: "no, we don't. We have none of that, either North or South of the Danube." After a long pause, during which I watched a mental procession of all the Greek women in my family, scattered across the islands and sitting there, as would become them, in the white shells of their stone houses, their eyes looking intently at the piece of cloth they were about to embroider, I said once again: "no, we don't". What else had the Greek women to do? Other than sweeping and cleaning the house, breeding a couple of goats and raising their children... there wasn't even any mud to fight with. Whereas the Aromanian women, perpetually on the road, always had to be prepared to lay down their house and then pack it up, to cook and cook for their folk and for the shepherds; they had to spin and reel and wind, to spool up, to sow, to paint, to knit, they had to carry things and carry ... and always carry something from some place to another and then back, with their fingers red with frost and cold water, clawed because of the spinning and knitting, carrying ill animals on their back. And if you want you can cross the Danube and have a look here as well. You go with the woman in the field, and there watch her hoeing, cropping, making hay stacks, then go back with her to the yard and stay with her while she feeds the pig, then the cow, while she cuts wood, spins, sews and cooks for a man who does not leave for some sea voyage, but stays home and eats three times a day. I am aware that, in doing justice to my Aromanian grandmothers and great-grandmothers, I am about to do wrong to my Greek ancestors, who were probably so keen on their useless frills, not because they had too much spare time, but because this was the proper thing to do in their world, which was so different from ours.

Pouqeville, who knew the Aromanians in the early 1800s, praised them for their moderation. Those who were seduced by the gaudy dress and gave in to the Albanians' questionable taste were held up to ridicule. In 1875, for instance, in Samarina, a ban was passed on the women's caps with silver ornaments, deemed to be too costly. The girls who refused to give it up were booed in the street. Gradually, the old trimmings were abandoned and their place was taken by equally expensive objects brought from the town.

It is not easy to do justice to someone in the Balkans. Generally speaking, you can hardly judge someone by their clothes: what is *too much* for some might be *barely enough* for others. Anyhow, it's worth remembering that the Aromanians made their clothes almost exclusively of wool and cotton, that the women did not make embroideries, that there were tailors among us, actually whole villages specialized in tailoring. Most tailors were male. There were also itinerant tailors. In summer time, when the sheep wool was spun and spooled, they went up in the mountains and, sitting under a leafy roof, they sewed clothes for the shepherds and their men.

Each village had a market place, and right in the middle grew an old tree. Under that tree *the masters* gathered on Sundays and holidays. That was where important decisions were taken. When the *cogeabas* came, everyone would rise to their feet, for he was the boss. Even the elders stood up, even though their venerable age made the others rise on every other occasion. Women were not even allowed to cross the market place. Their steps had to stick to the shade of the houses around. To dare and show your face in front of the men, in broad daylight, was a shameful thing to do. They could go to the water pump, though. There was nothing to be ashamed of there. The industrious ones left their buckets in a row and went on seeing to their business. Others took advantage and spent hours in small talk. A healthy society takes up gossip for draining purposes.

The place where the men gathered to talk was the coffee-house. It could be found in a special building or in one of the inn rooms. Men of all ages came there, there was the priest and the teacher, and also the gendarme; they read the newspaper, or tried to have someone who had been travelling tell his story. *"What's new?"* they asked, and the man told them. They played card games for cups of coffee. The loser paid. The postman also dropped by, after having sounded his trumpet all over the village. Not to go to the coffee-house at all was too peculiar, it meant you had something to hide; to go there too often wasn't good either, it meant you were not serious enough about your business. Everything had to be done with moderation, within certain limits that no one spoke of explicitly – yet everybody knew.

In the drawing rooms with walnut wood parquet flooring, brass manganese gives out a gleaming light. All around, piles of pillows and carpets, a mirror, chests covered in veal skin, an icon, photographs and sometimes the clock brought from America which strikes the hours all in vain, for a world that still counts time *a la turca*.

Legs squeezed tight under the black woolen dresses, the women busy themselves with the new guest. The oldest one starts asking him questions about his father, mother, his wife, children, brothers, brothers-in-law, nephews, friends, how they are these days – in case she knows them already, who they are and what their names are – when she learns about them for the first time. After the old woman receives her answers, the next in age takes the guest over, wants to know the same thing, only she asks for more details. As the protocol has it, a third, then a fourth woman follows, if there are four old women in the house. The next guest, in case there are more than one, undergoes the same treatment. This is how every piece of information, once it's been listened to twice or thrice, carefully considered, scrutinized and interpreted after the guest has left, becomes fixed in the old women's memory. *"The amazing thing,"* some guest would comment sometimes, *"is how well they know the age, name and occupation of relatives I myself can hardly say I know."*

I have taken over the description of a visit from Nenitescu. Personally, I have strong doubts about such a scenario. I am for instance quite skeptical about the old women's interest in the others' age, for those I knew found it hard even to remember their own age. Grandmother kept no record of the time she'd lived, yet used to say: *"so-and-so is a mother to me,"* which meant: she is older than me, or: *"I could be a mother to so-and-so,"* in case the person was younger. The life of my grandmother and of the old women I knew seemed to be like a stripe I could only measure by referring it to other stripe-lives, or by referring it to wars, to times of sojourn or times of travel, or to catastrophes. What is more, their way of asking questions presupposed a very oblique strategy. For grandmother, overt curiosity read lack of grace.

Just like a badly handled curtain, the red woolen cloth covered her face. Words faded away and her hands remained on her chest, crossed like in an icon. Marusa had died for her folks. Silver necklaces and braids gave a cold embrace to her frail body. The women dressed her up in funeral clothes, singing to every ribbon and flower with which they adorned her. Their song hid a muffled undertone, which was her throbbing cry. From the dark corner where the old women stood, a clawed finger reached out sometimes to mend something. When everything was ready, somebody brought the father-in-law's necklace, made up of fifteen *dubloni*, Marusa's age, for it had been fifteen times that the father-in-law made it known to them that he wanted her for a daughter-in-law.

The mournful singing and crying carried Marusa up the hill. One hundred men, all dressed up, took her all the way to Piatra Araua, where the bridegroom's procession had been waiting for them since dawn. On his white horse, the bridegroom rested his hands on the gun handles. His thin moustache was shivering like a butterfly's feelers. *"He looks like St. George,"* one of the aunts exclaimed just before leaving. Restraint, tension and a desire for alliance. One side loses and another wins, and that day's stake was Marusa.

When the Albanian *comitagii* threw themselves, in a stormy bloodthirsty pack, upon the wedding guests, they were, as Kavafis says, a solution. The red kerchief fluttering above her head like one single wing, Marusa joined the men and fought by their side. At sunset, the wedding party won. The men and women in the bride's and bridegroom's processions mounted their dead on the horses. Once begun, the wedding had to go on, although the bride left behind dead people and it was dead people that accompanied her on the way to her husband-to-be. Shortly after – it all happened around the 1820s – the two *falcari* were united and they all left Muzachia.

Now, this is what Marusa's story was really about. Old Dovana had a girl ready to get married, and he himself sent word to Tuna that he was willing to become in-laws with him. Tuna was glad to hear that, which means that Dovana was really looked up to, and had some of his men prepare the meeting where he went with gifts for the girl, a golden necklace, a fine expensive head kerchief and other things like that. The wedding date was also settled by word of mouth, sent back and forth over the mountains. In order to take the girl from the place they had agreed on, Hrista Tuna sent several hundred men to accompany his son, Lambi. And Dovana, too, took his daughter to the mentioned place, ahead of a big convoy. For the wedding they sacrificed 100 fat sheep and 200 lambs. Parson Papani did a lovely service where he sung in Greek. And while the wedding guests were celebrating and enjoying themselves, although there wasn't much drinking, for we are not really into drinking our heads off, they were taken unawares by a band led by Ali Pasha, who'd just pilfered governor Kurd Pasha's home in Ohrida. They were loaded with plunders, yet craved for more. They attacked the wedding party at four in the morning. The Aromanians got to their feet and dashed them real hard. Obviously Marusa had no business to get involved in the fight, she was standing at some distance. I just had her rise and fight to make the story more appealing... After the fight, both Dovana's men and Tuna's left Muzachia for good.

With weddings and funerals things are like this: whenever such events intervene, the wedding must go on, since an important thing, once begun, has to be completed. Otherwise it's like trying to stop the explosion of a bomb that has been set to explode. On the other hand, it may be that I've insisted too much on the spirit of death that gets hold of the bride during the wedding. Yet there are enough data that invite such an interpretation. In some places the custom was that one week before the wedding the girl was not allowed to eat, and during the wedding she was not allowed to talk, eat or drink. The bride had one right, which was actually a duty: to cry. Her head was covered with a thick woolen cloth, which would be gradually removed, one bit at a time. The wedding guests could only see her face after the wedding was done.

As to the bride's necklace, things are more complicated. There are many places where the necklace was offered together with the engagement ring. In the case of engagements made when the two were still children, the necklace was given "piece by piece", and every year the father of the bridegroom presented another part of it.

Engagements were sometimes concluded even before the children were born. Let's say that you and I get along well and decide to become in-laws. We shake hands. The only thing left for us to do is make children. If they give birth to babies of opposite sexes, the parents exchange a woolen diaper and then the father of the bridegroom sends one golden piece every year... but I've already spoken of that. What I find truly important is this: in case one of the children is blind, crippled or dumb, or if it so happens that one of the families grows poor, things go on, the wedding is held anyway, since breaking such an agreement would be an incredibly disgraceful thing to do. Yet such weddings were not merely a point of honor; as they were often made among members of different *falcari*, they functioned as a necessary circumstance for the exchange of women. If one *falcare* "spent" its women all by itself, its strength petered out. What helped me to fully understand the necessity of broadening the circle of those who marry among themselves was an answer Margaret Meed was lucky enough to receive from a young man who belonged to a very primitive tribe. She asked him why it was that he did not marry his own sister, and he answered: *"What a foolish thought! What do you mean, that I should remain without a brother-in-law? Who is to go hunting and fighting with me then?"*

I have another story about an unhappy wedding. A rich young man had been engaged for some years and now the wedding was drawing near. He then went to the fair with some friends of his to buy gifts for the bride and guests. When they reached the town, they went straight to the fair stalls that seemed richest in beautiful things. They could have stopped by the pub first, but what they had come there for was much more important. They picked their merchandise, negotiated the price and filled up a whole bundle of bags. They bought silver ornaments, Turkish head kerchiefs sewn with beads, mirrors and arms for the brothers and brothers-in-law, for the women young and old, candies, halvah, raisins, almonds, and sweet wines. They paid and left.

On the road people like to sing, what else can they do? So the young men were singing. When the highwaymen – Albanians, Greeks, Turks or Bulgarians – I have no idea what they were – hit, they fought with lions' prowess, but the rich young man who was about to get married was killed. They mounted him on his horseback and took him home.

When the father saw his son lying dead in the grass, he started to shout, but his wailing did not mourn the one who'd passed away, but the widowed and unmarried bride, who was to become his daughter-in-law. What's more, as if it hadn't been curious enough that the father was lamenting the daughter-in-law's sorrow more than he did his own son, the bride's father said: *"Don't sorrow over our sorrow, brother, you have your own to bemoan. Your son has taken the road with no return, our girl can always get married again. You still have your younger son, who's not married yet, you can give him our daughter to have as his lawfully wedded wife,* aman, aman, aman…*"*

There was heavy rain, and Rosca was marrying his daughter. There are certain memories which devouring oblivion cannot upset one bit. I wore plaits and a tergal dress I had received as a present in Greece. I was at their house, sitting together with my mother on the verandah. I still preserve the vivid image of the silky mud pools reflecting the outhouses and chattels in their yard. The bridegroom's procession came in running. They stopped a few yards from the verandah and took off their caps and hats, while a lanky old man from the bride's party stood up. It was raining and he started singing… *"aide more…"*

I cannot say much of the song other than it seemed to have no end. One by one, the bridegroom's guests stepped on to the verandah. The bride and bridegroom remained alone under the rain that soaked the starch-dressed aloysia flowers and made the girl's dress stick to her body, so that for one second she seemed to me to be naked and covered in a thin coat of plaster. *"Hurry up, Papu!"* someone tried to urge him, but the old man gave him a cold reptilian look and said in a whistling voice: *"We won't change our ways with the weather…"*

Unfortunately, the song lasted too long, and the bride had melted away. The old man's strong and knotty fingers grabbed what was left of her and trusted it with the women, for them to see what they could make of it. At any rate, several years after the wedding, the daughter-in-law grew fed-up with the mother-in-law and asked for a divorce. The wedding was all for nothing.

"With us, marriages to people from another kin could not happen, neither was divorce an accepted matter. It was the parents who settled the marriage for you and their word was not to be disobeyed. Love was of no consequence among our people." This

is what Caterina used to tell me; she was my cousin and elder by some thirty years. As I listened to her, I was thinking that I could not possibly put such things on paper, for we might be taken for a species of prize winners devoid of any passion or blemish. Yet, in order to draw a different image, I needed arguments. It was Caterina herself who provided me with one, although she did not really mean to. We were working on the family tree of the Safaricas. I was writing from Caterina's dictation. At some point I stopped to voice my surprise: "Is that so, grandpa Gheorghe was married twice?" "Yes." "The first wife died, she could not have children?" "No, she couldn't." "They had a boy together, but she had a limp in one leg. But did he not see that at the wedding?" "Forget it, leave it as it is."

Then I discovered that my grandfather's brother, too, was married twice. What's this all about, Caterina? Well, this is how it is, there's nothing we can do now... And don't you write about that, people will laugh at us!

And yet, here I am, writing about it, forgive me, Caterina! I wouldn't have probably written anything had I not discovered quite a number of similar cases, where the early engagement, made in childhood, was not respected, and the men fooled around and neglected their homes. I met with frivolous behavior in the women, too. The easy way of dealing with this is to believe that the norm was respected in times of old, while misconduct only appeared when the world started to disintegrate. No! I'd rather say that we were a society of normal people who lived by the rules, which some people broke from time to time.

Ali Pasha had spoiled the unity of Dumata's kin, they were too scattered now to count as one group. Hrista's father left, too. Wherever he went they called him the "Greek man", for he knew Greek. The rest of the family settled for a while in Moloviste, where the Greek man also came and put an end to his journeying. As he used to wear white clothes, people in Moloviste called him "Christmas man". Finally they all moved to Bitolia, where they tried to strike roots once again.

As for Hrista Dumata, although he was a serious and diligent young man, he suffered from consumption, and died. Death rushed to take him and nobody tried to stop it, as the hospital in Bitolia refused him any professional care because his nephews on his brother's side were going to the Romanian school. Hrista's last wish was to be buried in the Romanian graveyard where some others spent their eternal rest, among whom was leader Apostol Margarit and a certain gendarme from Pleasa.

The Romanian priest was late, Gopes was far away, and he had to ride his horse all the way down. On the other hand, though, two other people presented themselves at the dead man's house, with the exactness of a tax collector: the representative of the Patriarchal Seat and the Greek-maniac priest. On seeing them, Hrista's brother howled at them to get lost. He yelled at them until he was arrested. Then one of his sons and his wife, Constandina, threw themselves upon the priests to smash them. Madam Pinetta interfered then – she was Apostol Margarit's daughter. And she said some things, no one knows what exactly, for which she was insulted. Meanwhile, the gendarmes were hopping like locusts on either side of the dead man who'd been left without a candle. In the long run, Constandina seized the bishop by his beard and threw him down the stairs under everybody's eyes, while the people around were shouting their heads off, in all languages: *disari dispoti, oxo dispot, nadvor vladica, nafoara, afara* ... ["get out!"]

The fight continued for quite a while, but since the dead man started to stink, the two parties finally reached an agreement and buried him on neutral ground, which is to say neither with the Greeks nor with the Romanians, until the Turk, or else the pagan, had decided what was the best thing to do.

Nasu al Pariza is gone too. Well, he died, what can one do? At the dead man's house, there was bitter mourning. The women with disheveled hair, women from four generations, were moaning and calling his name, "why have you gone, why have you gone?" And the village people came to see him, one by one, all silent. Cups of black coffee opened the dead man's road to the other realm, and he lied on his back, dignified and as if upset. The clothes he wore were all white and new, and the silver of his buttons glistened in the room.

Time was passing by, and the priest had not come yet. The people were starting to fret with polite impatience. Few paid attention when the old man's bed was approached by his last daughter-in-law on the nephew's side, the one who'd taken care of him for several years. She was now looking at him with the angry hateful look of someone who'd been betrayed. She turned her broad back of a young black mare on the others, seized the coffin with both hands and started to jerk it, howling: *"why did you die, why did you die, my master, my* afendi*? You wanted bread, I baked you bread, you wanted meat, I cooked you meat, you had not one, not two, but ten purses of money you could count. Why did you die, why should we remain orphans?"* And smack! she slapped the dead man in the face twice. Then she turned to face the others and stared at them with an inquiring look. Since no one had anything to say, she walked away and went into the house.

I know this story from Gina.

Whoever is a *hagiu* has not lived in vain. Everybody believed that and it was his conviction, too. That's why when he set foot on the road to Jerusalem, he almost seemed to be flying; it was something that opened up paths for him. Now, on his return, his soul was blessed and his sacks packed full of talismans. The sun rose and set, rose and set, yet another tree, and yet another valley, and with every tree and every valley he was closer and closer to home.

He had several hours left to ride when his mind started to struggle with a bugging thought – the thought of death. If the brigands were to attack him, his people would be lost! He could not help turning his head back and all around, over and over again. When he actually caught sight of the bandits it was like a confirmation, see I was right! He spurred on his horse. The others kept close. He started to pray. While he did so he could hear them calling each other's names. God, help me, have mercy on us! Doubtlessly his words did reach God, for the bandits' bullets missed him and whistled away in the hot air. He prayed once again, even more ardently, God, help me and I will make a silver icon!

The wonder occurred instantaneously. God inspired the horse with such strength that his joints jerked up and he dashed off instantly, so grandpa found himself in the middle of the village, under the oak tree where the elders met to decide on matters of great consequence. People gathered there and started to comment on how much he seemed to have changed. The horse got a sort of dizziness which numbed him for the rest of his life, and Gheorghe the *Hagiu*, that is my grandfather, after he made the icon just as he had sworn to, and after he planted in my grandma's belly the seed of another son, that is my father, left for America and, within a few months, died.

Mother was the only one of us who had the privilege to see the icon concealed in a tree hollow like in a tabernacle. From what she told us I understand it is a piece of silver, just like a thick sheet of metal with a hole at head level through which the blackened figure of a saint can be seen. In order to be able to have a better look, she had to bend down into the hollow up to her waist and lit up a match to throw some light on that spot of the tree marrow which wears down my grandfather's sign of faith more and more every year – my grandfather who was chased by brigands sometime around 1905 or 1906, as he was returning home from his pilgrimage.

The talismans were taken good care of, yet they all disappeared in time, one by one, and so did the only photograph where, when I was a child, I could see a man wearing a kilt and leaning his left elbow against a column built out of seven rolls of cheese.

I cannot forget the point of argument that grew between my mother and father along the years, just as I was growing up. *"I am a general's son,"* daddy used to say in dead seriousness, counting on a story that went back to the vague past of family memory, one that told of a *celnic*, who may have been Manea Safarica himself, of whom Caragiani makes a note. *"A general of a chicken of a hatching hen,"* my mother replied, "so much am I an admiral's daughter…" After I finish this book, and I pray God to help me write it, I will get my paints and brushes and draw a portrait of grandpa Gheorghe. I will have him against a dark orange background, and he will wear a red belt, a white kilt and a blue vest with sides fluttering in the air. He will be armed with as many weapons as a man can carry, and his eyes and nostrils will breed fire, and under his arm-pit he will have a jar full of honey, inside which there will be the head of a bandit placed there for sweet conservation. Around him I will draw a whole bunch of chicken of a hatching hen cheeping cheerfully, and in the opposite corner I will do justice to grandpa Spiru, my mother's father, and draw him on the deck of a sailing boat. Of course he will wear admiral's clothes and will scrutinize the horizon with a pink spotty telescope. Above, instead of the sky, I will have an undulating green ribbon on which I will write in golden letters the names of my grandfathers.

We were arguing all the time, and our perpetual discontent came from a time gap. I was too young and she too old – in the tiny room in Mendeleev Street, we were like two foxes constrained to live together in a cage. It still shames me to remember how I was peeking into her small box with many divisions where she kept her funeral clothes, the candies, a round icon brought by her grandfather from the Holy Mountain, and the bills father gave her on holidays and which on the same occasions she shared with us, her nephews, in a random manner, since she had no notion of their value. I admit I couldn't help fumbling, and in doing so, I missed my chance to teach her a lesson, by unveiling her sneaky ways, the way she searched every little spot in the house. When there was only the two of us, I would harass her without mercy: *"why don't you ever ask an honest question, why don't you have the courage to speak your mind, why do you have to be so slippery?"* *"One needs to be diplomatic these days, girl,"* she answered, and this is the very word she used: *diplomatic.*

As to her methods of exploring the unknown, I still maintain today the point of view I held back then, since, leaving moral judgements aside, her attempts at satisfying her curiosity all by herself simply put her life in jeopardy. One day we came back home and found her nearly suffocated in the tiny space between the floor and the bottom of the cupboard, where she had crept in order to see what it was that mother kept in two cardboard boxes. All she had to do was ask me, who had searched the spot myself; I could have explained to her that it was the place where mother was hiding the Christmas tree ornaments from me. But she wouldn't accept that, she had to manage by herself, and that for the important reason that she needed to be diplomatic and because she felt that in our house, she was not, and could not be, the mistress. There was another thing, too, maybe the most important: in the very center of Bucharest, she still behaved as she used to in Doliani, where one did not want to ask too many questions unless one wanted to feel humiliated. The better attitude was to look carefully, to prowl around the object time and again, to see to what use the others put it, how much value they place on it, and only then, in case you were still unclear, could you ask someone else.

Otherwise, poor grandma would have tattled all the time. *"Shut your mouth, granny!"* father used to cut her short in the most brutal manner. To his lack of respect towards his mother, I answered with a similar lack of respect towards my father, and loved him with the anguish of a love that forbade lavish expression.

Grandma could easily remember the periods in her life when she'd been happy. The hard and sorrowful times spent in the Doliani house, the long and crooked years when she was a widow with six children, had been secluded in an iron-clad case which no one could ever break open. There was only one thing she agreed to tell me about; in the early years of her married life with grandpa Gheorghe, he brought home a big fish, but, you know, some fish, it was *this* big. In order to show me how big that fish was, she spread her arms like an old stork and tried to reach out on either side as much as she could. Yet, she could remember every single instant of the years she'd spent in Constanta, at the house of her most beloved and well-to-do son. "You don't like it here, with us," I'd say to her. "Of course I do…" "Did you like it too at uncle Cezar's?" That's where she used to stay for the summer, when we were off to the sheepfolds. "Chesaru is good too, yes, good…" "But you loved it better at Dimitrachi's, didn't' you?" "In Constanta I had my *cohea*." Which means that she had her nook from where she could keep a watchful eye over the whole house. "Sing something for me, grandma," I would ask sometimes, unaware that this request, "sing something for me", would become a key point of my profession. And she'd open a slit in the iron-clad case where the times of old lay hidden, and take out of it a lullaby, always the same, which she sang in a soft voice. Both the tune and her image as she sang it fold in my memory over the character in a Japanese film adaptation of *Macbeth* – a witch who mumbles an incantation while she spins a distaff in a hut overhung with spiders' webs. Grandma's song went like this: *"Moi, mamma will make silver buttons for you, mamma will make a* chilim *for you, more, for all the* picurari *and all the* carvanari *to lie down and sleep on it."* "Don't even think about it," I'd reply. "I don't need a carpet where all the shepherds and all the pit-men in the world could come and sleep. Since it would be mine, mine would be the toil, too, and try only to imagine what it would be like to have to work myself to death – waiting for them all at evening time in the porch, feeding them all, cleaning them and the kids…" Taken aback by my violent opposition, grandma finally spotted the misunderstanding, *carvanar* does not mean *carbunar* [pit-man], how could I have even thought of it? What business did we have with the pit-men, they were *farseroti*! The *chilim* would only welcome the shepherds and the *chirigii* [drivers of conveyances] from Veria, and grandpa was a *chirigiu* too, a *chiradzibas.* That was a spur to my curiosity. "What do you mean *chiradzibas*, what did he convey, where from and where to, and who paid him?" "Well, he just conveyed things, all kinds of things, of course he did, he was a *chiradzi,* don't you know?" And grandma would shut down like a book with pages that stick together because they were spread with marmalade.

And so no one could ever find out what grandpa Gheorghe used to convey. Father and Cezar had heard from Dimitrachi a story of how grandpa learned to make hard cheese: he hid in the attic and dug a hole in the floor in order to watch what the master cheese-maker did after he soaked the sweet cheese in boiling water. That is all they knew, and neither will I ever know what grandpa conveyed and why he stopped conveying when he did, who owned the caravan he led as a *chiradzibas,* how much they paid him, whether the cheese business was more comfortable or more lucrative, and if it was, why he left for America, which was actually all the worse for him, since after a few months he was taken ill with ague or hepatitis, and died!

Sweet cheese, the town tease. The first blocks of flats impressed no one. People were not one bit intimidated by the concrete colossi that pierced the night with rectangular looks. Gradually, though, the new neighborhood became a reality. And then the dwellers of the house-and-yards became furious, they thickened the labyrinth of lines where the laundry was hung for drying, they had their dogs barking, their children screaming, they fried meat rolls on coal and played backgammon wearing their pajamas in the hope that this way they could counteract the assault on their horizontal lives.

Yet the most spectacular act of defiance the old neighborhood was able to perform came from an old Aromanian called Dimcea Varghida, who ignored the new situation to the point of keeping undisturbed to his old shepherd's ways. The animals lived in a storehouse at the back of the yard – sheep that had long forgotten the taste of grass and had gone almost blind. They ate dry bread collected from the canteens. Once a year, an Astrakhan ram met their ladies' needs and, for one day, they could take joy in their weak curly kids. The next day the lamb was sacrificed for its skin and the little martyrs became a milk source for *bai* Dimcea Varghida and his family. Unwilling to put up with the small, shiny shit balls any longer, the old man's daughters would have preferred to grow flowers in the small garden, yet they were rather content with the gold trinkets their father bought for them. The sons grumbled when they had to carry the bread from the canteens, yet always had full pockets, and were less needy than the young men of their age. *Teta* Despa alone, his wife of a lifetime, supported him in his Sisyphan labor, and joined him in wheeling a huge white roll of cheese on the streets of Bucharest.

When they died, the sheep were sold. They turned the more meager ones into pastrami and the whole family was then invited to share the sin with them. I too was there and ate through my tears. Lying on covers in the yard, among pilings of roasted chops and tinder, we were eating in despair and hatred, each of us was gulping down the memory of the sheep herds our ancestors had. It was liquidation time.

My father's village is a stone's throw from the small town of Veria, and an hour and a half's ride from Salonic. That's where grandmother also lived; since she'd lost her husband in 1907, she had no one around, no one to help her, and took care of six children who hung on to her.

Here's a short version of a conversation I had with Sanga Vrana, a neighbor of ours in Doliani, who was about my father's age. *"Teta Sanga, who was there to help grandma, the folks on grandpa's side, or maybe her own family? I know she was related to rich people..."* *"We used to help each other a lot there, my girl, but you know... there wasn't really someone that close to her."* *"So what did she do for a living? Did she go to work for other people, or did she go to town?"*

The answer to such questions was a definite *no*. She used to spin, I know she did, and knit socks, but I don't suppose that's something to get you through feeding seven mouths. Some day, Sanga blurted it out: *"Dear, your grandma Caterina had a big two-storied stone house and got good money for renting it."* *"Rent it, you say, to whom?"* I rushed to ask. *"Well, to whoever wanted a place to stay. In summer time Aromanian folks from the valley villages used to come up here and rent a room or two, as many as they needed. And whoever had a house and needed money took them in."*

That was easy! I breathed out, relieved. My grandmother's existence was now conceivable and clear. She lived for 91 years. She liked to repeat that in her youthful years she used to be tall and beautiful, as beautiful as a mare from Ghimulzina! At the time I knew her she was limp, had a hunchback and had gone almost blind. I listened to what she said, yet could not believe her. When she died, though, her body relaxed and sort of slackened so that I could see how her head and heels pressed against the edges of the coffin. As she was lying there with two black kerchiefs wrapped around her head, she looked like a carbonized Madonna.

So one could live on rent payments… Days on end I tried to put together the image of a house occupied by tenants, who invade you with all their stuff and ways. I couldn't know whether they quarreled or whether they helped and spoke nicely to each other. I have the right to suppose both variants. Tolerance, or better said lenience, has had a long tradition in the Balkans, yet it can come to an end, and when it does, there's a burst of cruelty and the flowers of evil sprout up.

I am thinking that it was easier for the Aromanians to rent their houses, given that they had only recently taken to a sedentary life and therefore, in winter time, they had been tenants themselves in other people's houses on the plain. Such an experience makes you perceive the house less as a private space and rather as a temporary shelter you can make good use of. By repeated renting, they learned that a house can bring you money just as a sheep or a piece of land can. There's only one step to becoming an inn-keeper, which the Aromanians certainly took. Someone remembered once how he had come across a young Aromanian who had an inn in a very isolated and dangerous place. His family lived a peaceful life back home while he was struggling with unimaginable hardships. The man was perplexed, so he asked the Aromanian why he'd chosen such a hard way, and the young man replied: *"You see, sir, one can earn a living a lot easier in bad places, but where life is good, bread is much more hard to get, as everyone stuffs themselves on it."* They were inn-keepers first and then became owners of hotels and restaurants. Were not the Capsa brothers Aromanians, too?

The inns were usually very dirty places. All over the Balkans the inn was expected to ensure safety for the travelers and goods. Comfort counted for nothing in inns or, in most cases, in houses too.

Burileanu recounts how he was sheltered once in a house in Cerna village and had to sleep in the same room with six other people, a cow, a mare, a donkey, an ox and several hens. The only one who was thrown outside was the dog, for fear he might bite the traveler.

The photograph taken at my Christening features 58 men and women. Quite a crowd. Five of them are Greeks and fifty-three Aromanians. The proportion remained the same in my parents' entourage. Even the photographer was an Aromanian, his name was Papacostea. All I know about him is that he took good photos and that you had to trail him forever to ever get them. When the client came, whether he was a friend or not, he told him: *"you come today? They aren't ready, come tomorrow!"* The man came back the next day and Papacostea put on an angry face: *"why are you here today, I told you to come tomorrow, not today!"* The Aromanians had a really soft spot in their hearts for photography. The best photographers of all were the Manachia brothers. I will tell you their story.

He died of diabetes in 1964. His funeral ceremony was shot by a group of amateur filmmakers. Soon after that his wife, Vasilichia, and their son put on his grave a marble funeral stone that read: *"Tuca peciva Milton Manachia, 1880-1964. Prv filmski snimatel na Balcanot."*

So the lucky day of Balkan cinematography dawned in Avdela. But not in 1880, as one might think, but several years earlier, when Milton's elder brother, Yanachi, was born. Their father, Dimitri Manachia, first sent him to a prep school in Ianina and then to the Faculty of Fine Arts in Paris. Back from Paris, sometime before 1898 – yet by no means earlier than that – he infused his brother Milton with his newly acquired passion for photography. As a consequence, in 1898 they both opened a photographic studio in Ianina. The shop sign must have read "The Manachia Brothers". It was hugely successful. They were invited over to Bitolia all the time, which was the heart and core of Aromanian cultural life at the time. This went on for quite a while, and in 1902, they moved to Bitolia for good. It was all for the better! They opened a new photographic studio called *"Artistic foto studio"*. If I were to give credit to the article in *Frunza vlaha* ["The Vallachian Leaf"], of which I'm giving a summary here, although I am familiar with some of the serious books dedicated to the Manachia brothers by

Greek, Serb and Bulgarian authors, if I were then to take our people at their word, the Bitolia episode was a big success, too. They made portraits of the most important pashas, took photos of considerably or less considerably rich Aromanians. A more adventurous and romantic spirit, Maltu, that is Milton, took his camera and went to Avdela, Perivole and Samarina. He took photos of whatever chanced to fall under the eye of his camera.

In 1906 the two brothers participated in the first Exhibition of Balkan Photography held in Bucharest. They were awarded all sorts of prizes, among which the title of Photographers of the Romanian Royal Court. It was still in 1906 that Yanaki – who'd become a professor by now – came up with an idea: nothing more nor less than to go to London – *how do you do, how do you do* – and buy the camera no. 300. He did so and, coming back to his brother Maltu, said: *"Well?"*

Brother Maltu took the camera in his hands – I can imagine the emotion he felt – and started making films. He took a shot of grandma Despa, who'd reached the respectable age of 107. She was already around 102 when he took a photo of her in 1900, in Ianina. A bit of arithmetic will tell you that Despa was born somewhere around 1798, which is during the high times of Ali Pasha of Tepeleni... Maltu's films, will you bear that in mind, please, are among the first ethnographic documentaries in the world. I remember a *hora* that extends over several hills and valleys and also a number of images that portray the "torment of the wool", which follows its transformations all the way from the spinning to the weaving.

In 1911, the Manachia brothers took a photo of Sultan Mehmed Rashid V, who was then in Bitolia. Taking the photo of a Sultan! Every Romanian periodical of the time published the advertisement of the Manachia Studio, the sole bank of ethnographic images in the Balkans. I don't know how long Yanaki lived. In 1957 and 1961, Milton took photos of Tito. His last photographs were produced in 1963, when, at 83, he captured on film the wreckage caused by the earthquake in Skopje. He died one year later.

The Academy Library is in possession of a Manachia photography archive. We, at the Museum of the Romanian Peasant, also have some tens of examples. The richest archive seems to be somewhere in the former Yugoslavia. If I tried a little, I could find out where exactly. Word goes that they have some 6,000 negatives and

2,000 meters of film. I will never get to see them! The Manachia brothers are of interest to me to the extent that they are part of my list of "outstanding Aromanians". That they were outstanding is more than evident. Had they not been that notable, the Greeks would not have tried to have them pass for Greeks, the Serbs for Serbs, etc. They have written thick fine-quality books on them. But we don't need to bother, the Manachia brothers are Aromanians. As for myself, whenever I go through lists of names that feature Yanaki and Maltu, I recall the words Elenca Dudescu once said to her nephew, Ion Ghica, while she fed him quince jelly wrapped up in paper: *"...whenever I think of the nobleness of our family, look, it makes my head swim."* It makes my head swim, too, to think how good we were and what has become of us...

In 1809, Vallachian physicians could only receive the right of free practice on the recommendation of a committee which, in the said year, was made up of doctors called Gh. Schina, Silvestru Filitis, Constantin Darvari, Constantin Caracas and Emanuel Rizu. Which is to say Aromanians all through!

The first patient of the first royal hospital – Coltea – was also an Aromanian. He was Costin of Metovo, later known as The Crippled One; he was a *chirigiu* and on passing once through Bucharest, suffered the inaugural operation – the amputation of a foot chilblain. After he was healed, a royal grant offered him land to build a house on near Dimbovita, where he got married and became a rich man.

I know of no comprehensive study on the Aromanian physicians. Whoever might care to write one will certainly have a lot of things to put down. In 1706, Dimitrie Halchia got his doctor's diploma in Italy. Ioan Nicolide de Pindo (1737-1828) practiced medicine in Vienna, as the only physician in the Greek and Aromanian colony. In 1785, he was already a member of the Vienna Academy, and in 1794 he published a book entitled *Comment on the Advised Way to Cure the French Disease or the Malfrance*, written in vulgar Greek for the common use.

The Zagori region was the cradle of the medicine men called *caloiatri*, who were the inheritors of ancient medical traditions. They walked the fairs and shouted out loud for everyone to hear, look, here's the great doctor, the great *"hernier"* ["rupture-man"]. They even carried with them a sack full of "ruptures". They also removed cataracts, they cured all kind of diseases. Personally, I believe there were Aromanians among them as well. One argument in favor of this would be their professional argot. They called a house a *tufa*, a church an *aghiotufa*, a physician a *katafianos*... The Aromanians had a taste for secret idioms, and their performances as healers are a matter of common knowledge. It is said, for instance, that Cimu al Saputa of Gopes would remove bad teeth with his gun. Hard as I tried to figure out how he did it, I couldn't find an answer. I could partially understand, though, how Calciu al Dafin of Tirnovo put up his "gallows". They say he took a piece of wood, tied it with a rope at one end and placed it under his sole. He tied the other end of the rope to the tooth and now all you have to do is imagine what kind of person could that man be who, in one snatch of the head, managed to pull out his tooth.

Here is a story from Vodena which Papahagi learned from Gheorghe Celea. Negeb bey the Turk took a new shepherd's spool made of apple tree wood and put it on the coat of the sick man near the painful spot. In the scoop of the spool he placed a brazen coal onto which he laid a sheet of paper with some Turkish words written on it. He uttered several magic formulas and then had the man take off his clothes. Right on the spot from under the spool a small blister appeared which he covered with a plantain leaf. The blister burst and the man was healed. In the Cadrilater, another Turk applied the same treatment, and in Vodena a much craftier one healed men and cattle alike from a distance, without ever showing his face.

When someone was very sick and no one could tell what was wrong with him, several old women were bidden to his house where they spent the night sleeping; function of the dream they had they could "diagnose" the man and also indicate the place where healing oblations were to be performed for him.

He who was taken with vertigo had to eat roasted donkey liver.

Someone suffering from measles was not allowed to see fire, and in his house, painting and laundry washing was forbidden.

When someone had headaches, someone else spat on his forehead, and with a thread, took the measure of his head once and then once again; if on the second measurement "the head was bigger", it meant that the head "had opened up" and the man had to keep a cloth wrapped up around his head for a while.

For smallpox they had a vaccine: pus was taken from a sick man's festering wound and placed on a piece of paper or cloth and then a needle and thread was passed through the cloth and then through the skin of the patient.

Of the mad dogs, they said that they *had been taken away by the moon* or it was believed that they had eaten a lark that, after rising to the heavens, fell dead to the ground. It was also believed that it was enough to be barked at by a mad dog in order to go mad yourself. The puppies a bitch had on her first litter were not kept because they were prone to go mad. I cannot forget how Nacu's father, who was Pepo's husband, was bit by a mad dog and fell ill. For a while they kept the dog tied to the fence, but it became frantic and was in such pain that finally they had to shoot it.

I learned from Haciu's book how one could start a business without any capital. You could go to someone in the village who was a great merchant. He would ask you: What do you want to start, a grocery, a textile shop? Well, I fancy this. Good, here, take these wares and take this money, and tell me what else you need, then go to that fair. You took what he'd given you and then went where he'd sent you and thus became a "man of the trade". You started from here and could become a big shot.

The son of Stefan Cuimigiu of Bitolia, for instance, who became, after making a fortune, Coengiopol, was first a prentice boy in Parascheva Atanasi's shop. Then he became a great merchant, the head of all trade corporations in Bucharest, which is why he was among the delegates who received King Carol I in 1866. Coengiopol had a grand couture shop for gentlemen's clothes and also made ecclesiastic objects and sacerdotal attire. He is the founder of the national silk industry – mulberry trees, silkworms, fabrics. In Lyon he received the golden medal for Romanian silk. And he did all that in thirty years, because he was fifty-nine when he died.

Success comes in all sorts of shapes. For instance, Mihalache Buia, whom a *carvanar* of Gopes talked into coming to Bucharest when he was thirteen, ran the Macedonia Coffee-house between 1880 and 1924, then the Panziana, a finely decorated place lit by Japanese lanterns. He started business as director of the Vlasca Hotel, where he introduced a variety show... A daring move...

The next story tells of Jean Nicu, Panaioti's son, who was a silversmith in Greece. Jean's beginnings were fraught with poverty, his job was to carry firewood on his mule from Nevesca. He then had enough of it and went back to his father in Ianita, but I couldn't tell what he did there. He started out again to find his luck and went to Cavala, where, at that time, big fortunes were made and unmade. He was a poor man, so he found a job as a tobacco collector, unaware that he was born under the tobacco sign. There, in Cavala, he met one Max Popovici, who was in charge with tobacco brands' imports to Romania and who asked him, "don't you want to come and work for RMS?" "Yes I do," said Jean and he came to Bucharest sometime around 1892-1893. Yet here too perspectives proved too narrow for him and he accepted a specialist tobacconist's position in Finland. He then went to Stockholm as director of the tobacco supplies. How's that for a rising star? But there is more. Shortly he became general director of the tobacco companies in Norway, Sweden and Denmark. Then he took advantage of the closing down of the frontiers in WWI and bought the entire cargo of coffee and tobacco destined to reach Germany, which is to say he hit the jackpot with help from his fellow black-dealers in Venice. The contract stipulated his right to do business by himself, so he gave up his general directorship and started his own grand style businesses. And who do you think came and took his place? Spiru Odi of Moloviste, that is a "sympatriot".

There's always some spending to do when you have the money. How did Jean do it? First he had a villa built in Stockholm – a villa as white as a wedding cake, which he called Nevesca. Then he bought an island right in front of Stockholm, and after he had carefully husbanded his life, both on water and land, he started making offerings for Athens.

We've had a whole lot of wonderfully skilled silversmiths. So were, for instance, the Fila brothers; there were seven of them, as many as the days of the week: Dimitri, Adam, Nauni, Stavru, Iancu, Nicola and Gheorghe. They worked for the towns of Bitolia, Skopje, Alexandria, Salonic and Vidin. As the story goes, Gheorghe made once two candlesticks for the Sultan – but one had better call them wonders, not candlesticks. In 1860, the Sultan himself asked of him two pieces that should incorporate two date trees. The silversmith suggested a project that would represent two plane trees 6 meters high each, 31 candles, superb canaries' cages and basins with fish swimming in them. On top of each candlestick Fila carved an eagle. Beautiful!

There were master silversmiths and itinerant craftsmen. They specialized in filigree, an art that takes minuteness and rhythm. One could meet them in all sorts of places, particularly in Calaru and Seracu. Nevertheless, skill can mislead people sometimes. In Niculita, for instance, there was a nest of forgers, and at times, in the Cataflicu grottos spurious money was coined, too.

Everywhere in Epir, Macedonia and Albania, the river bridges were made by Aromanian builders and so were the churches, the temples, the palace-houses, and actually all important buildings.

Kanitz, a German ethnologist to whom we owe a lot, tells how in 1872 he met Nicolae Ficioglu, the builder of a grandiose bridge ordered by Midhat Pasha. He writes that from the way the man looked you couldn't tell him from an ordinary peasant and that what the "architect" was particularly enthusiastic about was the cost of the bridge, 700,000 *piastri*; money does count in the Balkans.

Haciu's book also speaks about the impressive sumptuousness of the houses. Niciota's house, for instance, counted 64 rooms, had been built in 1840, the façade was decorated with balconies and delicate colonnades, while the interior yard hosted several wells, in tune with the fashion in Triest.

Icon painters, stone masons and cutters... shepherds in the summer and craftsmen in the winter. The books on Balkan art refer to our Aromanians as Albanians, Bulgarians, Serbs, Greeks. I have this acute feeling that nothing can be done.

There have been among us monks, bishops, inspired priests, but frauds too. One such man, for instance, sold forged certificates testifying that some Christian girl willfully adopted the Moslem religion. The paper was needed for selling her to the Turkish harems.

It is also rumored that in Nevesca and Pisuderi, there were Aromanians who disguised themselves as priests. They put on the cassock and offered to serve the Bulgarians whom they Christened, married and buried while mumbling imaginary "services" which the Bulgarians could not understand, as they had no idea about the Greek language. In order to handle difficult situations, they had invented a "professional" argot that the mock psalm readers used in order to cue in the sham priests on some on-coming trouble. *"Open your eyes and see,"* the psalm reader sang with a nasal twang, *"beware the men in black and their nasty drumming!"* The "men in black" were the representatives of the Turkish authorities and the "nasty drumming" was the clubbing and whacking they were not shy to perform on someone's hind parts.

I have another story which I couldn't possibly leave aside either: the story of the priests "anointed" at Guva al Pardale from near Gramaticova. Guva was a cave where *lala* Gheorghe hid the young men who wanted to skip military service. He kept them there until he managed to buy priests' credentials for them. When the Albanian Aromanians of Gramaticova saw a priest who did not follow the churchly ritual norms, they figured it out instantly: this one, they said, went to the Guva al Pardale "school".

In a village near Gortcha, on a day no one can remember exactly, anyhow sometime around 1900, the villagers chased away the Greek priest and asked the Patriarch to send them an apt one who could do the service in Romanian. As the Patriarch's answer kept people waiting – it's a long way to Constantinople – what could they do? They sent a letter to the Pope in Rome informing him of their Latin origin and asking for a priest who was not a stranger to their language. A more prompt and efficient man, the Pope sent them a priest and so, within a month, he established an Aromanian parish of Catholic rite. I don't know why things turned out badly after all. Anyhow, the fact is mentioned in a newspaper.

A truly interesting story is that of the cases of Moslemization, a lot more numerous than you might think, on which the Aromanian scholars remained deftly silent, with the only exception of Nanta village, acknowledged to have turned to the church of Allah with all its members. Still, it's noteworthy that the case was accepted mainly because the greater part of the community was made up of *megleni*.

Moslemization occurred in a variety of circumstances. In 1790, 36 villages in Berat and Permeti decided to react to the extreme suffering the Turks inflicted on them by keeping a strictest Easter fast and praying... God, though, did not help them the way they expected. Then they called for a *hoge* and adopted the Turks' religion, while at the same time grabbing their weapons and taking, just like the Albanians, to a life of burglary.

Still in the 18th century, a family that were as down on their luck as a poor man can get, prayed to God and prayed to all the saints, and in the long run said to themselves, why don't we give it a try with Allah, and as luck had it, their life changed for the better. The whole village then followed in their steps and was converted to a religion that takes no candles to practice and where the salvation of the soul costs less. Once, a village went Moslem simply because the Orthodox priest was in the habit of coming late for the mass and thus kept the heads of the community waiting. Many villages were forced into adhering to Mohamed's religion by Ali Pasha, the satrap of Ianina, who sent an *imam* to them and one way or another the whole village, its priest included, turned Moslem. The lucky thing is that many such conversions only lasted for a short while.

It hasn't been easy for me to find these examples. Burileanu had the courage to face a reality others were too embarrassed to accept. He tells how he had in front of his eyes Aromanians who spoke Aromanian, yet were Moslems. In Spele, the Moslemized Aromanians met in secret to practice the Christian ritual and at home called one another by their Christian names. Mit Tase of Gradistea told Burileanu that they had in their village six families coming from Frasari who were *pasele* [pashas]. The Aromanian relatives of the families in Frasari were Bus Tase, Tamilici, Sin Dini, Steriu Colea, Petre Lici. There, he added, the Moslems Ismail Pasha, Adem Pasha and Alim Pasha were cousins of his great-grandfathers, and one of their sons, Malik Bey, was a rich man and he had his own song in Frasari and one could tell from that song how praised and respected he was.

There is a strange story telling of Hasan, the Moslemized Romanian who came to Burileanu with some business proposals. When they met at the hotel to settle the selling of some antique coins, he started talking in Italian. *Turco buono, turco fino, beve acqua, non beve vino.* Then he made a complete swerve, *turco buono, turco fino, non beve acqua, beve vino.* "You drink wine, you, a Turk? How come?" asked Burileanu. So Hasan Loci got to tell him of his folks, who were Romanians and had left Romania some 500 years before. They came there before the city of Tirana was raised and his family took part in Skanderberg's *Dieta*. As for himself, when he was rich, he traveled to Italy, France, Austria, Germany and Romania. Loci assured him that many Moslem Aromanians lived in Spata. By asking others about Hasan, Burileanu received confirmation that he had been rich and that his ruin made the fortune of 100 families in Elbasan and another 100 in Tirana.

Other examples testify to the fact that Moslemization was not always an easy process. The first action of Turkish propaganda was a lesson on ritual washing. Near Vodevita was a village where the Aromanian Christians would have no one lead them by the nose, so when Hogea tried to teach them how to wash themselves, they jumped in the water naked and shouted: *"that's how we do the washing, come and see how good it feels!"*

One day, Burileanu met an Albanian who spoke Aromanian. *"What are you?,"* he asked. The answer was blunt: *"Romanian."* *"How can you be a Romanian, since your parents are Albanian?"* I quote from the book: *"We used to be Albanians, sir, but lately we have turned Romanian, too."* It is said there that many Moslemized Albanians around the Frasari village still knew they had been Aromanians once. They were religious people, made the sign of the cross over the bread before they sliced it, and believed in Christ whom they called Azaret Issa. They called St. Nicolae Aider Baba. They respected neither the Ramadam nor the Bairam. God knows what was in their heads.

Someone might ask me whether this was all I had to say about the Aromanians' faith in God. In some way, yes. All the books that address this issue try to convey to us the admirable image of some perfect believers. The disconcerting thing for me was that the nomad Aromanians did not have a priest of their own, were not related to any church, or to any churchyard. Although I lack the valid arguments, I tend to believe that at some point they joined the bogomils' ranks. I spoke much to the same effect sometime in the '80s, at one of the George Murnu cultural meetings, and thus managed to cause a lot of vexation among the audience. One man who was present there wrote me some thirty pages where he described the religious life of the Aromanians of Doliani village, which happens to be the very place where my father was born. Out of consideration I did not reply to him, but I certainly felt like shouting back: Doliani village was founded in

1899, the question is what were things like one, two or three centuries before that time?

I'll end this episode with a little story that's as ambiguous as the whole affair. It tells of Vasilichia, the favorite wife of Ali Pasha of Ianina. She was an Aromanian and the Turk loved her madly. Some say she loved him, too, others say she didn't, but rather hated him bitterly, because on the day the satraps came to steal her from her family home she'd have sworn on her cross to get back at them someday. The remarkable thing seems to me to be the lover's gesture, who threw 40 *ocale* of sugar into the well whereof his lady used to drink water, in order to beg for sweet thoughts from her, and allowed her to have within the palace an altar where she could pray to her Christian God. After he died, some sources claim that Vasilichia retired into the peace of a nunnery, while others will have her end miserably in a brothel, a drunkard and opium addict. The choice is yours.

This is the story of the big black ram. One summer, some *farseroti* of Muzachia took their sheep to pasture on the border of a lake. The sheep were grazing when a big black ram came out of the lake. He was as tall as a horse, had horns the size of a cartwheel and, under his tail, paraded the immense regalia of his potency. He dashed into the herd, fixed all the sheep and, with a howling bellow, threw himself back into the water. The shepherds were little impressed with the event; they made jokes. They even shouted, "good for him", "that's a go", and things like that. But in the winter, when they returned to Muzachia, every sheep gave birth to two or three little lambs. They were black and had evil looks. A sort of bedazzlement got hold of the poor mothers, as they were jostled and shoved by some thick-tailed young rams. The shepherds turned a blind eye to the sign now, just as they had done before, so the next summer they returned to the same place near the lake, in all serenity. And while the sheep were grazing, what do you know, the big black ram came out again. He was now wearing a big silver bell hanging round his neck. The lambs, who had been frolicking by the side of their mothers, stole away, gathered around the ram's legs, went baa! baa! twice and followed him into the water. Words cannot describe the sheep's pain and the shepherds' fright. They never passed by that lake again.

Here is the matter Burileanu discussed with Tode Mitru, with whom he had just become friends. Tode asked him whether he was married. "Not yet, but I'm going to." "Do tell me when you get married, just let me know when it happens, I would very much like to be there." "But how will you manage to come to Italy?" Burileanu asked. "On mare back." "You can go on mare back all the way to Valona, but then you have a sea to cross." Mitu pondered for a second, slightly discouraged, but then insisted: "on mare back it'll be, don't you worry, I'll come on mare back."

The Aromanians traveled a lot and whoever travels a lot knows the code of the road:
- set out on Mondays or Thursdays;
- carry salt and incense in your pocket;
- do not go to sleep near fountains or under nut-trees;
- when the horse makes as if he does not want to move on, obey him;
- if you want someone not to come back, put salt in their shoes;
- if you want them to come back, make them step over a water bowl with a twist on it into which you have tucked a coin;
- the priest has to "read" the traveler's clothes;
- on the day someone is about to leave the house, nobody works and, most importantly, nobody sweeps;
- whoever spools up wool after midnight upsets the roads of the travelers;
- whoever spins after midnight spins the roads of those who are away from home.

The Balkan libraries have been burning for many centuries now. Only good fortune can keep a manuscript away from the flames. But if fire does not consume it, water will soak it, and if water does not soak it, an earthquake will bury it in the ground, and if it survives the earthquake too, it is sure to be brought to light by the wrong hands, which will either destroy it or indifferently put it back on a shelf. The chance a manuscript might have to be discovered the second time puts me in mind of an image advanced by the Zen doctrine as an illustration for the chances a soul has to be reincarnated. They say that on the bottom of the ocean lies a turtle that rises to the surface once in 600 years; in order for the soul to be reincarnated, the turtle has to emerge from under a ring that may happen to be floating anywhere on the ocean waters and to pass its head right through that ring.

Some writings keep to secret places, it's amazing to see how good they are at that. Just like Burileanu put it in 1905, in writing about the Ardenita Monastery: *"This Romanian monastery is now in the hands of the Greeks, just like every other monastery around here, and it has inside, as far as I learned, old documents, undoubtedly important ones, for some of them refer to the old Romanian villages that existed once in these places."* How many Romanians have searched the archives in Ardenita, who knows what lies hidden there? We have no clue.

This is another story taken from Burileanu: in Moscopole, after the catastrophe, those who stayed acted like they'd turned dumb or something. The merchants used the old books as covers for their slimy account records, and the big Gospel of St. Nicolae Church, which had escaped the arson – no one knows how – was left by the churchmen at the hands of the children who tied it with a stripe and dragged it along the street or used it as a prop to keep the door shut. Until, luckily, two Greek-maniac priests got angry and took it from them and locked it in an iron case. Was that case ever found?

In the 18th century the Aromanians used four alphabets for writing, and the history of these writing styles is sometimes quite amazing. For instance, as if it hadn't been surprising enough that in 1811 an Aromanian of Larisa printed an Italian grammar, we learn from an anonymous preface written in 1778 in Leipzig that that was a second edition and that the grammar had been drafted by Toma Dimitrie of Seatistea at the time when Theodor Cavaliotti was writing the first reader for the Aromanians. Anyway, what I find remarkable is that the first Aromanian writings are vocabularies and grammars, that is the instruments one needs in order to learn other languages, as if they hadn't known enough of them, their polyglossia being a matter of common knowledge.

Pouqueville claims that the Aromanians who traveled could speak a number of languages and that the book shelves in their homes were rich enough to entertain the traveler with refined tastes who was constrained to go around with only as many books as he could cram into his luggage. The Frenchman makes it clear that you could find excellent editions of the Greek classics, or of French literature, and the manner of his account is an indication that this kind of home library was not a rarity.

Although the Aromanians' love of books fills me with pride, I cannot possibly picture those who owned such treasures. They were neither professors nor writers, so I cannot understand where they found the time to read, how and when reading became a part of their active lives. And suppose they did read, say, the Greek classics, with whom did they discuss afterwards, and after they juggled with the subtleties of literary Greek, how could they go back to Aromanian? Did they not feel it was like a narrow coat they used to wear as children but was now an obstacle for the expression of grand ideas? But maybe I am wrong and it was not a coat, but a city, or if you dislike the city metaphor, say a mountain, a knife, a simple and dangerous weapon with a life of its own, just as any knife in Borges is. Aromanian was the knife you keep hidden under your coat, pressed against your chest.

The first one in our family to put the knife down was my father. As a punishment, he spoke poorly not only Aromanian, but every other language he knew. I can hardly understand the Aromanian written in books and Ana, my nephew, will never learn it, I'm sure. I wonder, how will we be able to talk with our ancestors on Doomsday?

What is an *Iradia*? It is a sort of Turkish ordinance. Why did the Sultan have to pass an Iradia in 1905 that would allow the Aromanians to have their church services performed in Aromanian and to learn Aromanian in schools? Because the pan-Hellenic movement considered that it was a sin to allow Aromanian in church and a crime to speak it in school. Much innocent blood was spilled for four decades on end.

Romanian education in the Balkans came into being in the second half of the 19th century, and started with a romantic story in which the main character is Averchie. Ioan Iaciu Buda, his father, and Tasa, his mother, who was a nephew of Hagi Ioti, himself a brother of Djuvara and in-laws with Badralexi, had no children. Therefore they started doing what people were in a habit of doing, that is they made ever renewed efforts to show their mercy and charity towards the poor, prayed, went to monasteries, built chapels. Finally they adopted a girl and God recompensed their efforts and blessed them with three children: Marita, Sanea and Atanase. Unfortunately the parents died soon and left the children to take care of one another. Time passed. In the winter they came down to Casandra with their sheep and then spent the summer up in the mountains at Badralexi's Calive near Salonic. When Atanase turned 61, an idea got hold of him. One day he asked for money from his sister and said he needed it to buy some sheep. She gave it to him and he went to the Athos Mountain to become a monk. When Marita found out, she dressed in black and mourned him as if he was dead. Her husband and *celnic* Badralexi, the father-in-law, sent a letter to the father superior of the monastery, threatening to go there and set them all on fire. The poor father sent the man home, he didn't need any trouble. Atanase went straight to Badralexi's home, not to his sister's, and told him about his plans. He talked about his faith, about his life's mission and convinced him. When they finished talking, the *celnic* sent two children to sister Marita's house to let her know that Atanase had come home. As she entered the room, she spoke through her tears: *"did you not think of our father's name when you went away to become a monk, had you no pity, or were you no longer able to live among people?"* We don't know what the errant brother answered. And somehow, what with this and that, several years later our man was in Bucharest, known by his monk's name, Averchie.

It happened one day when the newly formed troops were parading in front of Prince Cuza. A young man in a long cassock that hampered his movements was watching them with hungry eyes. For the first time in his life he saw an army that did not hoist the green Turkish flag with its crescent moon. Overwhelmed with happiness, Averchie shouted out: "I am a Romanian, too!" He had been sent to Bucharest by the Holy Mountain in order to settle things regarding some rights over the Radu Voda Monastery. He had plenty of time. So he joined a group of men who had initiated the fight for national awakening. He then took his nephew, Ioan Somu Tomescu, Marita's son, to the Central Seminary. The people Averchie was in touch with were called Magheru, Rosetti, Cristian Tell, Bolliac, Bolintineanu... As his new interests fully absorbed him, he neglected and thus spoiled his relationships with Ivir, the monastery on the Holy Mountain where he had taken vows. On the other hand, though, he received 4,000 lei from the charity trust, which helped him recruit ten young men aged between twelve and fourteen, all clever and learned. He also received, he alone knows how he did it, 20,000 lei to pay the teachers, the hostel, the clothes and food for a period of time deemed sufficient for the children to be brought to the level of the Romanian public schools. The ten young men will be the future apostles. And that's about all Averchie was meant to do.

Yet the Romanian language school really came into being thanks to Dimitrie Cozacovici. And this is another story. He was born in Aminciu. He graduated from an economic high-school in Budapest. In 1821 he was adjutant to Alexandru Duca. Four decades later, in 1860, together with other fellow patriots he organized the first Macedo-Romanian Committee in Bucharest. A founding member of the Romanian Academy, he willed his entire fortune to a Romanian school that was to be opened, after his death, in Tesalia or Epir. Fearing that an unwanted longevity might put off his design, he opted for an abrupt solution. He committed suicide! The executors of his will were V. Alecsandri, Cristian Tell, I. Bratianu and V. A. Urechea. He was also the author of the bilingual manifesto of the Macedo-Romanian Committee.

People were different at the time, they were easily inflamed. And so was Dimitrie Atanasescu, a tailor in Tirnovo, when he found in a Constantinople coffeehouse the manifesto of the Macedo-Romanian Committee. He was seized with passion. He threw the scissors away and started to go to school. He attended the courses of the Matei Basarab High-school in Bucharest, and in 1864 Prince Cuza himself helped him open a school in his native village. For four months everything went smoothly, but then the man was arrested as an ill doer. Alexandru Ghica put in a word for him, and Atanasescu opened another school in Bitolia, then returned to Tirnovo. He published many books among which *Tetrascript* and *Handbook for Young Boys*. Wide circulation, free copies. For special merits he received the Bene Merenti Prize 2nd class and a golden watch with nicely finished lids. Shortly after, he was removed from all his positions by the adepts of centralized education, since he was a representative of the old school, patronized by the county ephors.

Another story. According to my reckoning, Sima must have been born around 1780. She had a boy, Goul al Ghianci Cosmu, who married Mata, the daughter of Dori Craja who was a *chirigiu* and traveled from Gopes to Bucharest; it took him two months to do so. When they were rich, Sima dried the money bills on plates and spread them with a stick to prevent them from getting musty. But during Ali Pasha's time they became poor. Ghianci Cosmu, Mata's husband, was a tailor and money simply seemed to slip through his fingers; he tried to earn money for his family by traveling to Iasi and Constantinople. But all in vain! Yet the parents' travels can sometimes prove profitable for the sons who, sooner or later, follow in their steps.

In 1843, *chirigiu* Dori Craja's daughter, Mata, gave birth to Dimitri Gou Ghianci, who was later to call himself, after Cosmu, Cosmescu. He had several brothers and sisters. His biography was put together by the son of his brother Ghianachi who, in 1861, settled in Galati where he owned three bread ovens and the *Cazino* Coffee-house. One of his brothers joined him to help with the business and also to try and develop something for himself, then the father came too, then the tailor and finally Dimitri, who meanwhile had been working at a bakery in Bitolia. Business was fine for a while. Just that Dimitri was not really into pastry and tailoring, studying was his thing. And that was it, nothing more. A brother won't feed a brother, the nephew-biographer comments, but *"woe be to him that has none"*. So Dimitri received help from the others, who were hardly overjoyed, yet willing to do it, and he studied enough to become a teacher.

In 1865 he opened a school in Gopes, in one of the rooms of his house. There he introduced the methods of the Lancasterian school, the desk, the blackboard, and, most importantly, he inaugurated the model of the mixed school. An interesting detail the biographer mentions is the presence of one Sica, the first girl in Gopes who was able to read from the *Apostle*. An apostle himself and an exalted dreamer too, Dimitrie carried books with his own hands, struggled with the Turkish customs to pass them through, translated the carols edited by Anton Pann into Aromanian, translated the *Herods*, gathered and guided the first band to go carol-singing through Bitolia on Christmas Eve. Several years follow when we sort of lose track of him, anyway, he surely did a lot of organizing, initiating, managing and chance-taking. Suddenly, the trajectory of his life made a sudden down turn. Twilight of a beggared apostle – he spent the final part of his life in Ploiesti as manager of the Macedonia Hotel. Instead of numbers, the rooms were marked with the names of several villages that the owner had once taken to heart: Gopes, Moloviste, Megarova, Crusova. *"Give him the Gopes!,"* the owner told the "receptionist". In case the Aromanian guest did not have the money, the rule at the Macedonia Hotel was to offer him a free stay. Actually, the former apostle managed the business with his left hand, while with his right he handled the directorship of the Macedo-Romanian Committee in Ploiesti.

Would you like to know whether there is such a thing as a literature in Aromanian? Yes, there is. Whether it has a future? I don't know! It is very hard, if not impossible, to write literature in an idiom that lacks the support of some cultural institutions and state structures. Up to a certain point, the Aromanians used four alphabets for writing and spoke all the languages they came across. Their capacity to dissolve into other ethnic groups has diminished lately. The chameleons of the Balkans are shedding their skin. It all started towards the middle of the 19th century, when they began to claim acknowledgement for their Aromanianness. Sometime around 1900, an Aromanian innkeeper of Bitolia had three sorts of plates, on some of them was written "*pofta buna*" ["enjoy your meal"] in Greek, on others in Turkish, and on the third sort in Aromanian. A cosmopolitan and versatile spirit, in laying the table he took care to respond to the ethnic origin of his guest, and acknowledged the Aromanian his right of being an Aromanian.

Are you familiar with that strip of paper, often torn out of a notebook, which comes out from under your rug, slips in beneath your door, or pops up in the mail box, and which is a message from someone unknown to you addressed to St. Anton and to you at the same time – to him in the form of a prayer and to you as a sort of curse? It says, may your house burn in flames and your folks coil in pain in case you break the magic spell and do not add a link to this sickly chain. If I did not resent this horrible maneuver for a triple reason – first because it sneaks into my house like an insidious nasty bug, second because of its imperative style and third because, I don't know why, I tend to associate it with black magic – then I would indulge the delight I might take in a text that is obstinately copied over and over again. And because I like to believe that there is no system of self-regulation at work here, I imagine that, at some point, the curse would have been addressed to St. Anton and the prayer to me, and then, the text would have gradually changed into a sonnet, a romantic song, a geometry problem, a Kabalistic sign … anything.

The comparison might seem far-fetched to you, what follows has nothing to do with St. Anton. I'm simply going to talk about several poems I read in an Aromanian newspaper. The lines are associated to a perfectly obscure name, Mehmed Said, whom the paper editor presents as a Turkish colleague he once met at the University in Liegru. And, he goes on, as the Turk was once in Macedonia, he had the chance to listen to some Aromanian songs which he enjoyed very much, so out of a romantic zeal, he collected them, or rather had someone else put them down and translate them into Turkish for him, and then he himself translated them into French. The editor received the lines in the French variant and decided to give a Romanian version he then published. So here we go, Aromanian, Turkish, French, Romanian… I took the liberty to intervene here and there, since I thought that after such an adventure, anything goes.

The son of the pasha of Salona is by the river,
Reining his black horse, a restive stallion,
When Despa comes with her bucket to the water.
But from up there in the mountains, Anton watches
With an eagle eye, and he's her fiancée!
The son of the pasha of Salona gives a kiss
To a bunch of flowers and throws it in the river,
The water floats it to Despa. Despa picks it up and
Tucks it in her red belt.
But Anton rubs the flint steel of his gun,
Aman, aman.
Despa sings sweet charms to her flower,
"If only I knew Anton picked you up,
Precious flowers, sweet scented breath,
I'd put you in a glass at night next to my bed;
If only I knew Anton tied you in a bunch,
Precious flowers, smooth water-dew,
I'd carry you close to my bosom till you fade;
If only I knew Anton threw you in the river,
In the path of water that's running on to me,
I'd hang you by the icon and ask mother
To pin you in my hair when I am gone."

Anton trains his gun, not on the pasha's son,
Nor on his black, fiery horse,
But right on Despa's belt where
The dismal flowers hang.
Ay, ay, ay, Garcia Lorca would say here.
And he trains his gun again, not on the head,
Nor on the heart of the pasha's son,
But right on his right hand,
His cursed hand,
So he can never again pick up
Accursed bunches of flowers.

Amazing how many ethnic groups found shelter and an identity in the Balkan Peninsula! And if they managed to survive in such a small place, that is because they practiced the art of cohabitation. Sailors and merchants here, agricultural workers there, somewhere else mercenaries, each people defined a vocation for itself. The Aromanian alone cuts the figure of a universal man: besides navigation, he makes his presence felt everywhere else, he is a shepherd, a convey driver, a smith working on gold, silver, copper, stone and iron; he cures illnesses, fights, builds houses, is a weaver and a knitter, he digs out coal and bakes pastry, prints books, animates the commercial centers in the Balkans, fills the mountains and the valleys with his sheep. Fuelled by this tendency to extend a mastering embrace over as many things and places as possible, the Aromanian plays the roles of several characters. The transparency and simple solidity of his inner being, doubled by an extraordinary energy, help him bear the burden of so many disguises.

Someone who has a taste for such debates over origins as that asking what was there first, the egg or the hen, can safely state that, with the Aromanians, the sheep was there first. Stepping in the tracks of the herds, the people learned to spend the winter on the plain and the summer in the mountains, and for the rest of the time to live on the road.

Similarly to the shepherds' way of life, the power rests all with one person, every individual has to know how to do everything that needs to be done. The knowledge of the group is not divided, but re-emerges in its entirety within the person of every member of that group. And whatever it is that he knows how to do for himself and his own people, is something he will be able to do, at any moment, for others too, when he goes to settle in a town.

But is it right, I wonder, to imagine that once, in a distant past, the Aromanians used to be an enormous mass of shepherds who lived all in a heap with their sheep and that at some point, a second phase of their history began when they dismounted their horses, first only partially, in order to move their herds from one pasture to another, and then altogether, in order to live in built houses? Or is it more likely to suppose that the Aromanians lived, from the very beginning, a sedentary life, which they abandoned, due to the dire shifts of history, for a life on the road? Both hypotheses seem unconvincing to me. I believe that "from the very beginning", which is to say from the first centuries of the second half of the first millenium, they lived at the same time nomadic, half-moving or sedentary lives, that they "dwelled" in the mountains and on the plains, but also in villages and towns – that, in other words, you could find them everywhere.

I am, probably, the last person to attempt an informed description of the villages and towns inhabited by the Aromanians, so all I'll venture to do is present some pieces of information I found in Athanasie Haciu's as well as others' books. Haciu lists the villages and towns by "fronts", he thus writes of such fronts as Aspropotam Pind, the lower Epir, Epir-Tesalia, Albania-Adriatic... A fortunate choice of words, this *front*, for it really was like a war!

When the moment came for me to decide which way I was to organize all that information, I tried looking for a model that might suggest the autumn leaves falling down on the pavement. I couldn't find one, so I finally opted for the most disagreeable ordering I know of, which is the alphabetic one. Therefore, here we go, starting, not with Moscopole, but with A.

AMINCIU

We call the place Aminciu, but the Greeks call it Metovo. As for its origins, Haciu says that it was the Aromanians who founded it, and he also mentions a rather bizarre social phenomenon typical of the area. The distinctive sign that differentiated the three classes – the nobles (*cogeabasi*), the merchants and craftsmen and, finally, the mob (*populu*) – in everyday life was a red string they wore with their coat. If a person was a noble, the string was sewn around the collar and descended all the way down to the coat tail; a merchant or craftsman could only have their string down to their waist; while the *populu* wore coats on which the sole red embroidery was around the collar. The *populu* were not to be seen in the central market place. They had their own field outside the town instead. Marriages between "upper" and "lower" persons were inconceivable and to me it seems only natural they were so, since I can imagine the dilemma facing the child of such a misalliance if he had to decide on the appropriate length of his string.

Metovo is the birthplace of Averoff, the *everghet* who divided his fortune and willed it to the people of four towns: Metovo, Vurtunosu, Cutufleni and Ameru. His generosity reminds me of the founder of the *Osmanli* dynasty, the first Osman who was so open-hearted that he walked the streets of Brusa and shouted, "whoever is hungry, whoever is thirsty, whoever needs anything, let them come to my house." Averoff did the same, but the people of Aminciu put on airs in return, and when his nephew Evdochia, a big *evergheta* herself, came by, they forbade her to enter the town unless she put on a traditional suit. The poor woman was wearing European clothes. Anyway, if you happen to leaf through a peasant dress picture album and find the beautiful garments of Metovo, remember that they are Aromanian, and so is the attire of the *evzeni* who walk with a proud gait in front of the Parliament in Athens.

I couldn't go further without telling the beautiful story of *celnic* Floca. Once upon a time, a vizier who had a feud with the Sultan found shelter under the white coat of *celnic* Floca. A gentle soul, the *celnic* welcomed him, gave him food and drinks, in the winter had the vizier accompany him wherever he went, and on St. Gheorghe's Day, took him up into the mountains, to Aminciu. There was not much for the

Turk to do, he just dallied around, and from time to time told Floca how grateful he was and how he was going to manifest his gratitude some day. Floca was not easy to impress, and set little store by his words. But the time came when the wheel turned and the Vizier was back in favor. As a sign of recognition, he drew up a pronouncement for Floca whereby the Sublime Porte allowed the Aromanians of Aminciu to take their sheep to pasture freely both in Tesalia and Epir, exempted the village from the presence of the Turkish authorities and, what's even more, required the Turks who happened to pass through Aminciu to shoe their horses anew as they went out of the village, so that they should not even deprive the people of Aminciu of the dust on their roads. Taxes were slender and, most importantly, Aminciu was allowed to offer sanctuary to runaways, just like a monastery or city fortress. The people of the village lived on to enjoy these favors until the day when Ali Pasha of Ianina, the satrap, the tyrant, and the enemy, got his claws into them and denied them every right they had earned in Floca's time, which made the people leave their homes and go into the wide world.

The word also goes that when Floca received the pronouncement, the *cadi* of the neighboring town read it and then summoned his *zapcii* and told them, "take care not to bother the people of Aminciu." One of the *zapcii* wondered then, how were they to tell the people of Aminciu from other folk? "Well," the *cadi* replied, "you will keep an eye on the crossroads and prick your ears. If you hear idle talk, those will surely be people of Samarina, it's OK to catch them; if there is talk of cattle-theft, you're sure to deal with people of Perivole, the arch-thieves, so you bust them right away; and if you hear people talking of beautiful women, leave them alone, they are Floca's people of Aminciu, the Sultan is on their side!"

ARBANASI

Jirecek claims that Arbanasi and Moscopole were founded at the same time by people of the same kin, who'd come from Epir. They were rich people and could speak both Aromanian and Greek. Big business with Hungary, Poland, Russia and, of course, with the Romanian Principalities, as they were the first Aromanians to come on a large scale to Bucharest. At any rate, Iorga took them to be the "Greeks" of the commercial companies based in Sibiu and Brasov. Possibly. Information on these companies can be found in Olga Cicancici. I for one wanted to know more and went to Sibiu to see the churchyard where many of them were buried. Poor Radu had to wait for me for three hours, sitting on a funeral stone (luckily the sun was shining), and I copied down all the names written on the crosses. When I was back in Bucharest I got my purse stolen right away and so the whole documentation was gone...

The women of Arbanasi were famous for their good looks. Kanitz says that, when you stepped into the house of a rich Aromanian, you could swear you were in one of the castles in Northern Italy. He had no reason to lie about that. And the most handsome house, according to Kanitz, belonged to a certain Brincoveanu family. Incidentally, Prince Constantin Brincoveanu deposited his money in Venetia with an Aromanian banker.

In 1797, the town of Arbanasi went to rack and ruin. As a consequence – Haciu notes quoting Caragiani who quotes Jirecek – its townspeople flew to the four winds. Many of them came to the Principalities. I wonder, though, whether they were not planning to anyhow. And I will ask myself this question every time I refer to catastrophes of the sort, such as the burning down of Moscopole, the burning down of Gramosta, and so on; because I will not have it that serious, rich and healthy people can ever allow a foolish Turk to throw them out of their homes as he well pleases. I should mention that a village called Arbanasi also exists somewhere near Buzau. I'll call that pure coincidence, until someone investigates the matter.

I have no idea how many dwellers Arbanasi had in its times of glory. It's hard to figure that out, since the Turks did not count the women and children. The men paid an annual tribute (*haraciu*). Their heads were measured with a sort of bridle. They also paid the tax that exempted them from military service. There were no censuses.

AVDELA

Some say that Avdela goes as far back as the Byzantine Empire. Here is what Haciu writes of it in his book:

"A great displacement of the celnici *of Avdela occurred under the leadership of the grand chief Badralexi (Barda Alexi) who, together with 600 armed families withdrew to the south-east towards the Veria and Neagusta Mountains, where they founded the villages known by the name of Lower Selia, which included the old Selia, Marusia and Volada, and repopulated the town of Xirolivad (Uzungiova) which the Greeks had deserted. The exodus might have taken place sometime between 1775 and 1821."*

Anyway, since Badralexi's Calive have been brought up, I should add that among the Aromanians who were there from the beginning, besides the Avdela people, there were also dwellers of Perivole and Samarina, as well as a small number of *batuti*. The list of the tax payers of the Avela of Pind village includes, besides names like Paciurea, Cutroci, Dimu Buru, Tocica, Bagua, Dauti, Damasioti, Exarhu, Iani Buru, Iani Barda, Gogu Hagi-Simu, Caracosta, Bucuvala, Ceapar, Papahagi, Manachia, Saguna, Tanasoca, Busulenga, Cialera, Cavachi – someone who might be my first recorded ancestor, known as Mane Safarica.

And what do you think Caragiani's father, the professor of Iasi, was? A *caravanar* of Avdelia who had been working since 1850 on the Avdela-Bucharest route. He was the one who, in 1865, drove the 10 Aromanian boys whom Averchie took to the St. Apostles' School, one of them being Tuliu Taci, son of *chirigibas* Georgachi Taci, who worked on the Durato-Valon-Corita-Constantinople-Larisa route, and who talked business directly with the governors of the Ottoman provinces.

When priest Cosma passed through Avdela – he was in charge with the Greek propaganda, and instituted the Greek law in 40 villages of Zagor – he was beaten up and chased away, because the Avdela people were not afraid of a man with a long coat and couldn't care less about his curses.

What was Avdela's contribution to our renown? Let's see – fighters: Nicolo Giuvara, Alexi Traguda, Panaiot Benachi, Captain Dipla, Caciandoni's man; scholars: as I said, Caragiani, the dynasty of the five Papahagi (Pericle, Nicolae, Tache, Valeriu and Marian), then Nusi Tuliu, who had the gift of a great poet, T. Caciona and the two founders of the Aromanian school, Nicolae Tacit and Apostol Margarit, who opened at least as many schools as Spiru Haret did, but no one seems to remember that.

BITOLIA

The word goes that the people of Moscopole had started coming to Bitolia even before their town was torn apart. I have this feeling that in the Balkans one town is raised when another is about to fall. It seems that in order for Bitolia and Skopje to flourish, it took the downfall of Seres and Perlep, plus the Moscopole people I mentioned, among whom were the Nicarusi, Goga, Busa, and Sulira families, all of them very rich people. When the Nicarusis married their son, the engagement badge they sent to the girl was a plateful of Austrian golden coins. His houses had pavilions with greenhouses where rare plants grew and canaries sang. Not that Nicarusi was the only rich man around. Once, at an auction where the *Valiu* of Bitolia was the auctioneer, Ghiorghi Ciomu, a supplier of the Turkish armies, requested the concession of all passers-by, so he was asked who was to vouch for him, and Ciomu answered, "the porters". "What porters?" the *valiu* asked, angered. "Well, the very ones who are to come and take the money bags away from here."

Sometime in the 1890s, Faveyral estimated that in Bitolia there were 15,500 Aromanians. Born in the diocese of Lyon in 1817, the head of the Lazarian monks of Bitolia had a special sort of consideration for the Aromanians. Nenitescu describes their encounters in lavish detail; I'll give here some instances, which seem to me to have particular poignancy. The Frenchman felt so close to our people that he used to say, had he believed in metempsychosis, he most probably would have been, in some prior life, an Aromanian. He showed Nenitescu a drawer full of manuscripts and explained to him that, after he was no more, the Aromanians were to find there "everything that concerns them". He believed the Aromanians to be endowed with gifts no other Balkan people possessed and that the Vallachian people would play a role in the unification of Christ's churches, now at variance with each other. Nenitescu tried to take a photo of him, but the monk refused. He died in 1893 and, what do you know, Nenitescu received from an Aromanian of Bitolia a photo of Faveyral lying in state, which he introduced as an illustration on page 257 of his book. The poor old man could no longer protect his face after he was dead. I learned about Faveyral from a member of the Hasoti family, who owed him the French he could speak.

In the town where, on Easter night, the priest hailed "Christ is risen!" in six languages, the church had an important economic role to play! St. Dumitru, which was a rich parish, functioned as a crediting institution for the merchants of Bitolia, Vlahoclisura, Castoria and Magarova. You're not going to believe this, but the St. Dumitru Church even credited the Turks! The people placed their savings with the guardians at up to 9 percent interest, which was something. And the church, an important owner herself, issued paper money called *cahmei* worth one *piastru*, five or ten *parale*, and these bills were circulated even as far as Nijopole, Bucova, and Magarova. Relationships among people were based on trust. Ioan Sonte, for instance, who kept his commercial registers in Aromanian written with Greek characters, was appointed church guardian and general cashier for life, and this was an incredibly exhausting job with no payment at all. This happened sometime in the late 1800s.

After 1900, the town turned progressively to the Greek law, yet Haciu reports that nowhere else did he find a town-hall seal inscribed *Olah muhtari*. And the birth certificate (*nufuz*) of an Aromanian had clearly and legibly written on it the word *Romanian*. Bitolia was the birthplace of Dumba, an Aulic councilor in Vienna and an awfully rich man.

CALARU

"...Seen from the Buru (Buros) mountain, it seems lost at the bottom of a crater," writes Haciu. *Buru* is the name of one of the oldest Aromanian families attested by Caragiani. *Burr* is the Albanian for *man.* Calaru made a fortune out of dealings with Venice. I can think of no better place for the story of Tal Simota than this section, so listen! Tal Simota came from Vlahoclisura and her sole relation to Calaru boiled down to one bowl. A wine decanter of the sort people used to call *"bucale"* or *"birbile"* or *"bojane"* and which all around the year lay still on the shelf in the better room, and on holidays alone was brought down to do honor to important guests. It was actually glazed earthenware, which Pericle Papahagi tended to believe was of Viennese origin; as an argument in favor of his theory he invokes old Atanase Simota who, as early as the middle of the 18th century, had opened a workshop in Vienna. No one knows whether the decanter broke or still lies unnoticed in some house or museum, but we can surely tell what it looked like, from a picture published by Papahagi in one of the issues of the Academy Annals. An earthenware expert told me, after examining the photo, that it looked more like an Italian product, but since she couldn't lay hands on the actual object, she couldn't bet on it. So I give up on the origin of the bowl. But there is a whole story, or better said poem, to this bowl, which stirred Papahagi's interest as well. On the potbelly, within an oval medallion, lies written in Greek letters the following stanza in Aromanian:

> Călăreţu ameu, biá vinu ca pi ateu.
> Multu se nu biae, se nu te vemai.
> Tra se nu ţi se fache reu, tra se nu te mbetu eu.
> Vnă oară se biai şi acase ţi se vai.

> (My dear rider, stay awhile and drink
> But don't drink much or you'll be sick.
> Too much drink won't do you good,
> So have some wine and then go home.)

The poet seems to have been more ethical than he was inspired. Pericle Papahagi discovered the glazed earthenware sometime around 1900 and all he could learn from old Simota, and I believe there is no offense in my calling her that, given that she had by now reached the venerable age of 95, so all he could learn from her was that the bowl had been in their house for almost one hundred years.

What's the story with this quatrain, you may well be curious to know? The inscribed objects have had a long history and have made quite a career in the Balkans. It is said that in Bitule (Bitolia), Zisi Goga engraved on knives and yataghans lines that he himself composed in several languages. You went to Zisi whenever you needed something more original or personalized – since you could find objects inscribed with good wishes and smart sayings in fairs – and told him what it was you desired. You paid, he carved.

The truly breathtaking experience was, I believe, for Papahagi to set eyes on the twin copy of the bowl, on which, within an identical frame, was written a quatrain, this time in Greek. Here is what the lines said:

> May all horse-riders live long
> In good health and joy
> And drink the sweet red wine
> That pleases the heart.

An encomiastic formula, therefore, of the kind written on the beer jars and wine decanters to be found among the Germans, French, Italians, Austrians and so on. Such being the case, I find the text in Aromanian all the more interesting. Papahagi analyzed it from an orthographic point of view and found that the author had learned how to write within a linguistic system that was superior to the one practiced in Moscopole. The complex prosody is another element arguing for the existence of a cultured, circumstantial sort of poetry. We suppose that an inhabitant of Calari ordered two identical bowls, one without an inscription, which could be added at a later date – which is a sign that some local workshops might have existed. And we cannot opt out the existence of such workshops, given that, as early as the late 13th century, the bills and testaments in the area recorded such (I would say) Aromanian names as: Pietro Greco detto Cimilarca or Mathius ditus

Peliza. Doesn't that make a case? Maybe it does. But I'll ask the question differently: could not Atanase Simota, who had a workshop in Vienna, or maybe someone else like him, have had a workshop in Greece as well?

At any rate, what's certain is that the Simotas' bowls are evidence that some sort of "literary" life existed in the Aromanian small towns sometime around 1800.

CASTORIA

A small town that produced 200 millionaires in a jiffy. The money was made out of fur shreds, sable fur and nails, small marten skins, tiny pieces that never exceeded four millimeters, which were sent home by those who were traveling around the world, doing business, in places you least expected. And what do you think the people of Castoria did with those leather pieces, which were too small even to taw? They sewed them together with small stitches – a technique even children could learn in school– and manufactured fur "planks" 60 to 80 cm long that worked just fine as padding for the winter coats.

Castoria is very close to the place called The Handsome Oak Trees, where David, Sisaman's son, was murdered by some *Othites* Vallachians, which is to say, probably by some Aromanian *caravanari* – a deed that the Aromanian historians have always been shy to mention, reckoning that it's not civil to begin your history with a murder.

Anyhow, the people of Castoria themselves proved quite uncivil for centuries on end, as they refused to offer shelter to the people of Moscopole who were down and out by now. And to make things even uglier, in order to bar the Moscopole people's way to their town, they asked assistance from Kiazim Pasha, that is a Turk. I'm not trying to write a history of Aromanian infamies, but I will not turn my head away when there are any coming down my road.

CORITA

Aromanians from all over the world streamed in to fortify the town: shepherds from Gramosta, craftsmen from Moscopole, Sipisca and Niceea, silversmiths from Crusovo, tailors from Pleasa.

They all piled in and mingled so much that, if you asked a child on the street what kind of Aromanian he was, he may well have been unable to tell you.

CRUSOVA

As the story goes, it once happened that some Serbs entered the town of Crusova and said, *"Ovde sfi su veliki gazdi"*, which is to say, *"they are all rich people here."* At its best, Crusova counted 3,500 houses. When Weigand saw it, he exclaimed: *"This is Venice on dry land! It's marvelous!"*

Crusova, too, fed on the drifting Moscopole people who reached it sometime after 1764. The people came in groups, the priests leading, and built up neighborhoods. Fourteen neighborhoods were thus raised. Then, I couldn't tell when, the Gramoste people came too, among whom Papasotir Papasteriu's family, into which they say that no more nor less than eighteen generations of church people were born. The last one served for his whole life in a church in Piatra Neamt. His ancestor, one widow Zoga, came to Crusova, or so they say, with a capital of twelve thousand sheep. She was a sort of Tal Celea, if you remember the story in the first part of this book.

Voyvod Pitu Guli was also born in Crusova; he was a voyvod of *comitagii*, and died in the 1903 fire started by the Turks who were sworn enemies of his own gang. That's when Niciota's house burnt down too, that palace of 64 rooms that's been so much talked about.

As to the people of Crusova's way of life, I can tell you, for instance, that around 1915-1920, Gheorghe Bandu sang Aromanian songs with a harmonium accompaniment. He brought the instrument with him to Bucharest and sold it ten years ago.

In reading about the Crusova people, I learned they practiced a particular sort of job, that of *politagiu*, which was a kind of *carvanar* who conveyed valuables, money, and correspondence. That they were rich people is a fact evidenced, for instance, by the kingly demonstration of Hagi Busu, who helped the Turks with 900 horses. They say that after he did it, he buried his treasure in a place no one knew about, and then died.

Another craft the Crusova people were good at was cattle fattening, which they practiced especially on rams and male goats. The male goats were good for tallow and pastrami. In the autumn, one family alone could butcher as many as 4,000 cattle. Imagine how much blood could be spilt at such times! One male goat could weigh up to 70-80 kg, no less than a healthy grown-up man. The cattle were butchered in three sessions: first, some twelve thousand for tallow and bones, then they sacrificed the animals for the pastrami, and third, the rest walked away on their feet for two months or so, until they ended up in the form of chops to fill the bellies of those they were destined to in the first place. Thus, for instance, Ghica Totili took both his and other people's sheep to pasture in Volos, Patras or Pireu. Sometimes he herded up to fifteen flocks, each of them as large as 1,500 heads. He had under his orders seventy or eighty shepherds. The mutton sort called *civargic* was in great demand with the Turkish cuisine. The wool was sent to London, the tallow I don't know where, and they used it to make soap and candles, and what did not go into soap and candles was pressed for whatever was left there to be used for making oils.

They distilled turpentine, kerosene and mastic out of tree resin, made ropes, woolen coats, did tailoring, watch-making, and had a copper foundry... Haciu writes: *"Professor Vanciu Cocu has informed us that one professor N. Carascachi of Crusova, based in Larisa, who just died in Volo, would have written a complete monograph on Crusova, which is unfortunately lost."* Won't you believe how they get lost, these manuscripts on the Aromanians, lost like they never were! Despina was about two years old when, under the helpless eyes of my mother, she tore to pieces Stere Hagigogu's manuscript, which was, or so they say, a synthesis on the Aromanians of Verieni.

Anyway, it seems that the people of Crusova were keen to keep up one custom, for wherever they went, and they did travel a lot, it went without saying that on St. Mary's Day, dead or alive, they had to be back home.

FRASARI

I have nothing to tell you about the place once known as Frasari village, I'd rather we talked of the Frasari people, or *farseroti*. Caragiani said it before me: it cannot be that all the *farseroti* were natives of Frasari, since there were, as Caragiani estimates, around 200,000 of them, and the village could only shelter several thousand. It is, above all, an etymological conundrum. One suggestion has been that the ethnonym *farserot* was derived from the name of Pharsala – an unrealistic hypothesis, which did not fail to bring together Frasari with Cesar and Pompeii. Linguists had better work harder on that, what they've come up with so far is highly unconvincing.

Leaving etymologies aside, although I too have a theory that would send us all the way back to the *oseti* in the Caucasus, the first thing to say is that the *farseroti* were the most warlike and, up to a certain moment, most conservative branch of our kin. If someone chose to wear a garment other than the traditional one, anyone else had the right to tear it off; when a young woman married a Greek soldier, her brother came back from America with the sole intention to kill her. I can't believe it really was like that; it would have been terrible.

I am absolutely determined to consider the *sulioti* Aromanians of the *farserot* branch. I could almost say that unless you read about the Republic of Suli, you have lived in vain. The *suliot* women fought like Amazons, and after the downfall of the Sultan, Foti Geavela's mother was hired in the French army as a major. In Romania, the *farseroti* became good agricultural workers. In America they were the first to establish mutual aid societies. Wherever they might be, though, their true vocation is artistic: painters, sculptors, actors, singers, cinematographers.

FURCA

The *celnici* of Furca... Some *celnici*, upon my word! In the late 1800s, Ghiciu owned, hear this: 8,000 sheep, 5,000 goats, 4,000 mules, 200 shepherds, and no one even tried to count the dogs. Well, they say that Dimitri Ghiciu once went to Constantinople with some business and entered the palace court on horseback. And the Sultan humored him, didn't say anything like *move away*, although the governor was all worked up and in a quarreling mood. On his departure, Dimitri was offered a retinue that would aid him on the road and add a touch of splendor to his march. Ghiciu had four other sons of the same sort, and also had his son-in-law Ianuli, an awfully rich man, just as the Sufleri and Darda families were incredibly well off, too.

The Furca people were, first and foremost, tailors. Starting with 1875, many of them came to Vallachia, through the Danubian ports and Teleorman, where they started trading in cereals. And, you might like to know this: the Furca people claim that Karagheorghe Petrovici's grandfather or great-grandfather was an Aromanian, born in their village; Karagheorghe Petrovici is the founder of the Serb dynasty. Furca was also the native place of dear old Constantin Exarhu, the man who struggled for several decades in order for us to have the Athenee in Bucharest, and whose descendent has a radio show called *The Night Shift*, a young man of delightful wit endowed with a sparkling sense of humor.

GOPES

An old stratum of Meglenoromanians and on top of it a newer one of Moscopolean extraction. The Gopes migrations have been studied. Whoever finds the time to read will be amazed at the rigorous way in which the ones who were about to leave prepared the arrival of those who might come next.

GRAMOSTE

It is said somewhere in Haciu's book that Gramoste was an inexhaustible source of Aromanians. This is how I came to picture the Gramoste Mountain like a sheephill. Actually, the village was surrounded by a round of eighteen mountains on which our sheep grazed the grass centimeter by centimeter.

Famous *celnici*: Hagisteriu, Paciurea! Here's something to wonder at: *celnic* Paciurea built a *milkduct*, a glazed earthen supply main that was several kilometers long and which carried the milk from the sheepfolds all the way down. They bred horses for the men's pride. The man could say: *I own 200 horses* and that meant something.

The big barracks on the outskirts of Bitolia were raised by *celnic* Hagi Steriu. The town governor once sought counsel from him and kept complaining, not much of a rock supply, not much money either, but there couldn't be a garrison without a barrack... and so on. Hagi Steriu cut him short and said: *"Get on your horse and show me the place where you'd like your barracks to be built."* *"Here,"* the Turk said and the *cipan celnic* lifted a rock from the ground and told the Turk: *"Now you lift one yourself."* Those were the foundation rocks. The Aromanian closed the business with the words: *"From now on it's my business."*

A truly terrible sight must have been that of the burning down of Gramoste. Here is the story of doctor Filip Misea, as reproduced in Haciu's book. It all happened on August 15th, on St. Mary's Day. The year was 1760. The town was celebrating the patron of the church. A dozen priests were officiating in front of a crowd of nicely dressed people who were praying, heads down. Everyone was singing. I believe it was beautiful. That's when the two signs occurred: the miracle working icon cracked loudly and the icon lamp that was burning under it died away. I can hear the folklorist saying: *It is a topos.* I will tell that folklorist: *No shit!* The people were scared, what can that be?! And at the same moment two *caravanari* burst into the church, and, barely catching their breath, cried out that *a big Albanian army is coming to get us!*

Two men started to toll the bells, bang! bang! The people prepared themselves for the fight in no time and, God knows how the fight might have turned out – I see streams of blood gushing from all sides – but the heavens did not want that to happen and a heavy rain started to pour, or, what am I saying, not a rain, a flood... The waters of the Nicola rose, swollen with rain, so that the Albanians who were on the other bank had nothing else to do but to sit down and wait for the disaster to run its course. It seems they had to wait for a long time. Meanwhile, the Gramoste people mounted whatever they could grab on their horses and... ran for it.

When the waters receded and the Albanians could finally cross the Nicola, no one was left to menace with the yataghans, the town was deserted. So they could only perform the usual fire setting, pilfering and breaking, which turned the whole affair into petty thieving. The improper – to put it mildly – thing the Gramosteni people did – which the tradition preserved – was to go off their heads with all that hurry and abandon the miracle working icon that had warned them in the church.

Starting with 1760, the people of Gramoste spread all over the place: Livadzi, Zanita, Coceani, Nijopole, Magarova, Crusova, Blata, Hrupistea, etc. One fourth of the 40,000 Aromanians came back after a while, but got scattered again because of Ali Pasha's assaults. I will remind you, as I have before, that the Aromanians prefer to associate the quittance of a place with a catastrophic event. It may be that the Gramoste people would have left anyway, the temptation was too strong not to.

Another story of Gramoste. Once upon a time, they say, there was a *valama*, that is to say a horse guard. The best of them took care of the wild brood mares. One day, he was on his way, herding the mares, when he reached a mine, somewhere near Volvo, and there he found a big pile of gold and silver coins. It happened in 1750. He loaded the treasure onto his horse and went away with a light heart, heading for the house of *celnic* Paciurea, the man he worked for, and asked him whether he cared for some golden and silver horse hoofs. That's as much as he could think of.

According to G. Zaverca, the young men fought almost naked, handling catapults; giant bones were dug out of the cemetery. The women were famous for their strength. They got off their mules, gave birth and then mounted again. Some, they say, wore their hair short and put on men's clothes. And who could forget the story of the young girl for whose sake a whole *falcare* retreated in the Caragiova Mountains, lest the Turk should steal her and take her to his harem. Or the story of Sana, Hagiu Steriu's daughter... The stories are many and painful.

IANINA

In the 13th century our sheep were already grazing your pastures and it took the Venetians only one century to turn you into an important center. Iorga says that in the 17th century you were as large a town as Marseille. Do you admit that Maruti's and Ghioni's colleges, which brought you so much fame, were financed by the Aromanians? And what else haven't we done for you, and in return, you gave shelter to Ali Pasha...

And you, Ali Pasha, you wretched miserable man, did you forget how you fought the Turkish pashas of Andrutu Verusiu Muciana's army and how you begged our people to get for you, because they had the means to, the position of a *dervengiu pasha*? They trusted you, probably thinking of the story of that vizier who'd won Floca's protection, and you too may have claimed you were going to respect their privileges, and then, what did you do? We did wrong to fight by your side against Ibrahim Pasha and the pasha of Epir, 'cause the next thing you knew, you'd become a pasha of Tesalia, it was in 1784, and they also gave you the title of Rumili Valisi, but Hughes claims, and the man knows what he's talking about, that you used fake documents in order to obtain those titles... And even after you'd done so, Vlahava, Bucuvala and Hristachi helped you to beat the *bajbuzuc* rebels. They should have left you at their hands, pest be on you!

And then what did you do? You did some sharing out, didn't you, have this one, Veli, and you this one, Muhtar, and that will be yours, Salik, so all your sons, those raven babes, had their share of our villages, and we were no good now, mere slaves you loved to see begging and rambling and wandering from place to place. I will never forgive you for the dirty job you did when you crushed the *sulioti*: treason, betrayal and juggling. But has anyone ever told you how monk Samuel walked on the battlefield, the Bible in one hand and the sword in the other, or what he did when your people got to the gunpowder in the basement of the church and were about to blow it up while he was inside, do you know that he shouted out, "set that fire!", which is to say he still had the last word? Does Samuel never visit you when you're asleep at night?

You were friends with a whole lot of assassins, sneaks, perjurers, thieves, and, how strange is that, quite a bunch of good people too. Because, actually, not even you, bastard, lacked good qualities altogether. You tore away from the Porte, worked on your own, and shook off the moths from the banners of the Crescent Moon. At a time dominated by anarchy, you tried to instate order by means of anarchy. Napoleon used to call you "my respectable friend", and sent you Pouqueville as a gift, and we, the Aromanians, owe him so much, we would be so much poorer without his books. I wonder why it was that you supported the Greek letters, and encouraged crazed Cosma, who bullied everything that could be bullied into adopting the Greek law? One year before you died you understood how much you'd lost by pushing away our people. You tried then to make amends, but it was too late. Hursid Pasha stifled you.

Let me give the names of some of those who helped you: Nutu of Capesova, Buroglu of Doliani, Puliu Scrima of Perivoli were members of your administration council; Gheorghe Turture of Saracu was the minister of commerce, A. Sarmaniotti was your general officer of the fairs, and Colovos of Metovo your secretary. Ion Coletti, Turture's nephew, was your medical doctor, secretary and ambassador, Nicolo Mihu Pisca administrated your estates in Tesalia, G. Ceaprasli of Saracu was the personal doctor of your son Vali Pasha. I won't say much of the people under arms, I'll only mention Karaiscaki.

You slain us, you persecuted us, you burnt down our towns, and all the same, in the long run it was still us you could count on, and had you trusted us all the way, who knows what history would have looked like. And now, to round it off, I have something terrible to tell you: after I have read six books about your life and picked up all sorts of information from others, I came to the conclusion that you were a Moslemized Aromanian, one with a corrupted soul and madness gnawing at his heart, so that you struck us with the fury one can only lay bare in striking a brother, for this is the only way I can understand why Vasilichia loved you the way she did: being an Aromanian and a woman, she must have sensed something.

LIVADZI

Aromanians from Gramoste, for whom Ali Pasha was too obnoxious to live with, people from Moscopole and Perivoli and other less important branches. I've noted down a legend *celnic* Ioan Cepi told Haciu, a story that might explain why the Gramoste people ever came to stay in Livadzi. *Celnic* Cutuflesa had a very beautiful daughter and Ali Pasha had taken a fancy to her. I think things happened like this: Ali ordered that someone go and get the most beautiful girl so-and-so had. In the case I'm talking about, the father and some of her brothers left for Ianina in a hurry, in order to ask the tyrant to bear with them a while, until Easter time. Ali understood they had to keep their fast and said, all right, hardly suspecting the prank they were going to play on him. Cutuflesa went home, told everyone how things stood and, time to move on! They gathered all their sheep, everything they had, and left. Miss Cutuflesa, who'd just barely made it, was left with a shepherd whose name was too beautiful to be true, they say he was called Birbilu. Birbilu was entrusted with taking her to Casandra. But Ali Pasha's men, who were not born yesterday either, found out what they were up to and crossed their path. They all started fighting in the middle of a herd of 40,000 sheep. Ali's men on one side, on the other the shepherd and *celnici* Gheorghe Stapator and Tegu Paputa Arosa al Barba, who wore red shoes which nobody ever saw, as they were covered with golden coins. The Turks stole Tegu's sister-in-law and asked for a ransom. Many lives were lost there, on both sides. So the Aromanians gave up Casandra and looked for another place to settle down. There's no word left on how Miss Cutuflesa was recuperated after all.

MULOVISTE

The people of Muloviste traded in leeches. They did trade in other things too, but the leeches were their thing. They fell into two categories: first quality leeches, which were caught by Bulgarians diving naked into the swamps, and second quality leeches, which were collected off the skin of the oxen. The merchants only traveled at night, and in the daytime kept their leeches in bags of water in the shade.

The little slimy things were sold in Serbia, Austria, Germany, and France. It was the time when medicine often resorted to leeches. This is what Dimcea Cionu, doctor Tascu Trifon's uncle, dealt in back in the 1820s. He did business with France, Germany and Hungary, whereto he transported leeches, rose ointment from Kazanlic, aqua fortis, rice, tobacco and edible oil. Another merchandise he carried was saffron, in grain or ground. The demand was great.

There were 12 itinerant priests in Muloviste, who performed religious services in a number of villages. How they did it is what the next story is about. Once, one of the priests was called to serve at a Bulgarian's funeral. He couldn't go, so he told the family: here's what we'll do, at this o'clock you put him in his grave and rest assured that at precisely that time I will be singing the funeral service for him at my place. So said, so done!

Muloviste was the native place of Belimace, Velo and my goddaughter Alice's mother, Toma's mother and Jean's wife, who live in the Moghioros Plaza.

MOSCOPOLE

Let us first have a look at Haciu's book. He says that opinions vary as to the century that saw the settlement of the town: Aravantianos claims it was the 16th, Pouqueville the 11th. A 1903 issue of *Convorbiri Literare* ["Literary Conversations"] invokes a document in which an allegation is made that after 1453, Roman elements of Constantinople..., well, to make a long story short, it claims that in the wake of the downfall of Byzantium, the inhabitants of Constantinople were in the process of settling a considerable number of towns. Hahn attributes the success of Moscopole to these refugees. Anyway, since it was during the second half of the 15th century that a number of Balkan settlements began to strike roots, the hypothesis of a Constantinople transplant is worth our attention. That the Aromanians existed prior to this migration seems to be attested by a document preserved at a monastery in the vicinity of Corita, which is referred to in a 1930 Greek magazine. It is said there that Ioan Nicolau of Linotope built the monastery of Zermas in 1164, at his own expenses. Or, Corita, Linotope and Moscopole were situated, so to say, *dans le meme coin.*

Capidan, who was an Aromanian of Pind, had the heart to leave personal biases aside and claimed that the Aromanian branch in Albania descended from the Gramoste people and that it was them who peopled Moscopole. There is a tale in Borges talking of the same uneasiness that comes with an excessive loyalty. However, Capidan did not exclude the existence of other Aromanian groups either.

According to tradition, Moscopole must have had 72 churches, some 12,000 houses and a population of approximately 70,000 souls. The river that crosses the town is named Dzega, which I have also encountered as a family name with the farseroti. *Some say there were, at a certain moment, 300 workshops and 14 corporations in Moscopole. Women wore expensive clothes made of atlases, velvets, or brocades, adorned themselves with Persian shawls, with golden and silver ornaments, and swept the floors with brooms bedecked with Turkish gold coins. As for the eating habits, their table manners were complete, they used gold and silver cutlery and Italian faience plates, since Italy was not far, and neither was Venice, so they had Venetian mirrors. Otherwise, the Aromanians were a sober and prudent sort of people, but they were sometimes enticed by refined and precious things.*

Before the downfall, Haciu's book continues, the Moscopole merchants had the pashas of the Balkans at their beck and call, and managed to snatch privilege after privilege from the Porte. But their luxury was a sort of *hybris* that stirred the anger of Ali. I've told you already, I don't believe the first thing about the story of the Great Fire of Moscopole. Burileanu says that at a certain moment, 12 of the mayors of Moscopole – whom I identify as 12 *celnici* ruling over 12 town areas – killed one another inside the church of Prodrom. It all started from a financial conflict. Such frictions seem to have existed in other towns as well. The times were awful. It was not easy to do commerce with Venice under the Turks' nose. The world was simmering, the bands of *bajbuzuci* were on the rampage. The Russian and Turkish War and the Peloponesian revolts broke out. What were the people of Moscopole doing at that time?

At one particular moment the idea came up of sending the Provost of the Academy to disseminate the Albanian version of the New Testament among the Moslemized Albanians. The missionary was not successful, since being a Moslem was worth all the privilege. They tried this and that, eventually realizing that they wasted their time there. What's more, commerce with Italy was disastrous. Signs of good omen were surfacing, however, in the North. I will now copy a fragment of the Chronicle of Prodrom – one of the monasteries of Moscopole – which gives a succinct version of the facts:

"In the year 1769, no one contributed any money because this same year they (the Turks and the Albanians) came, looted the monastery and the town, while many houses were burnt down."

The document of Budapest, which I have already mentioned, testifies to the same attacks of the Turks and Albanians. There were two of them. And guess who was the leader of one of these attacks? Was it Ali? No, it was his father. Of the other attack, the one of 1788, they say there wasn't one stone left unturned, as a matter of fact, all the region was devastated: Sipisca, Bitcuchi, Niculina, Linotope, and Niceea. This is taken from a note in Haciu's book:

"On the walls of the monastery of Saint John the Baptist, which was surrounded by forests of fir-trees and pines, one could read, prior to the final destruction, the following words : 'Alas, Moscopole! Where is your beauty, where is the splendor that made your renown before 1769?' 'The wicked and the looters are the cause of my downfall! ' 'May the name of God and of John the Baptist help you restore the beauty of your earlier days!'"

A recurrent question that comes to my mind is whether the destruction of Moscopole and of practically the entire region would or would not have been possible, had there been the will to resist on the part of the people in power. I share the opinion of those who think that the powerful and the rich had already deserted the region by that time.

Dusan Popovici claims to have made the acquaintance of one family that worshiped a rod, "Moses' rod", which the leader of the community carried along during the exodus. They could also remember tales of pots filled with golden coins covered with bee-honey that were presented as victuals for the children, in case the Turks inquired about their contents. And they remembered more, of coats that were padded with money bills and credit notes – the "good clothes" meant to keep them "warm" wherever they went.

This is what I say the ending of the story of Moscopole was – it is a hypothesis I cannot support with actual data, you either believe me or you don't. I think that Moscopole was burnt down by the Turks at a moment when nothing mattered any more. It was Valeriu Papahagi who brought me round to this point of view. Who can imagine, after all, a vulture that stays for the rest of its life near the chinks of the hatched eggshells it came out of? As business with Italy went down, Moscopole lost the limelight, too. For centuries on end, the Aromanians had followed their sheep herds, it was now time for them to follow the hoards of the round yellow coins that could be profitably invested in Budapest or Vienna. The gates of those cities opened wide, the Aromanians were well received and, consumed with their own welfare, were completely assimilated within just two generations.

If one is in favor of the famous story of the town that was burnt down by the Turks, then one will also encourage the image of the inhabitant of Moscopole as a defeated man. This I will not have. I do not believe that the stock Aromanian is a sentimental person, but he can carry his native land as an inward dimension that can be neatly wrapped up and carried around wherever he goes. The poor were left behind in Moscopole, because they had nothing more to lose, while the Western cities had nothing to offer them. They were not up to the challenge of salvaging the town from the hands of the Turks. There is a legend, yet legends are hardly reliable, which records an exodus that started out from Moscopole and ran its course in several waves. It says that, during those hard times, there was a *celnic* famous for his wisdom who hanged three butchered hens in the market place of the town. One of them had its feathers plucked altogether, another partially and the third only here and there. From now on sources are contradictory, some say that the *celnic* left a note at the place of the hanging, while some others claim that he went there in person and, while pointing at the poor devils, said to the audience: "Behold these hens, brothers! Those of you who hurry to leave the town will be like this hen that still keeps its feathers, which is to say they won't lose much, they'll practically lose nothing. Those of you who linger and stay behind will end up like the half-plucked hen, which is to say they will lose much but there'll be some left for them to cling to. Those of you, however, who stay, for the sake of their houses, churches, printing presses, schools and for the sake of other things that we put together in such a short time, will end up like this wretched thing, this naked hen, and alas for them. This is all I had to say!"

Therefore, to me, Moscopole is the stage of a drama, but not of a tragedy. In seeking the truth, Valeriu Papahagi has not searched the walls and legends, but the shadowy vaults of the archives, and I think he found it.

NEVESCA

There is a story about Nevesca, too, and it goes like this: nine brothers set out to Bitolia to call the banns for a tenth brother, who was to marry a young and probably rich woman. While the boy was at the parents' house, asking the girl in marriage, the others eloped with Nevesca and took her to a neighboring forest, where there were some shepherds and some thieves who would not let them stay or pass. In the fight the girl was murdered, and this is why the community that settled there now bears her name.

The primitive stratum of that settlement was made up of Gramoste people. Next came the people of Moscopole and Nicolce, and finally the *farseroti*. Everything went well until Ali Pasha and the pest struck them and the wandering began once more. Around 1760, a community of Aromanians, lead by the Tarlea family, arrived in Alba County, others went to Tracia, and some others to Muloviste, as did the Paligora family of the saffron merchant.

And, please, tell me now, how was it possible for the inhabitants of Nevesca to migrate in 1760, when the people of Moscopole couldn't have come there before 1769, i.e. before the time of the first great fire?! Why should the Moscopole people have settled in Nevesca if the place was so bad the Nevesca people themselves had to leave it? I can only suppose that the people simply moved on, driven by interests that can no longer be reconstructed today.

Some say that the married women of Nevesca would curse a man named Ceanana – who was later to be known as Ceanescu in the Principalities – because it was him who first took the road north of the Danube, and many of their husbands followed in his steps and left home. Mihalache Buia, for instance, who owned the *Pariziana* garden, had lent his ear to the smooth talk of a *chirigiu* of Gopes and had ended up in Bucharest. When the *chirigii* and their caravans went through the villages, they exerted on those with a mind to distant places the attraction circuses have elsewhere.

OHRIDA

Well rooted at the junction of two main roads, the one leading to Ianina and the other being the European road to Constantinople, this town was founded by Samuel, the tyrant of Serbia, or this is what Pouqueville says. Ohrida was situated in the vicinity of the spot where the Aromanians murdered David, the son of Sisman, who is said to have been granted a golden bull by Vasile III the *Bulgarocton*, whereby they both had become Christian subjects of the *Exarhat* of Ohrida. Haciu claims the whole thing is attested in a manuscript by an Aromanian priest, dated 1205, which is preserved in a monastery in Ohrida.

Here too came former inhabitants of Moscopole, after 1800. Good stock people and all. For example, the brothers Iancu and Atanasie Dimonie. They owned a caravan of approximately 200 to 300 horses. One day, thieves in the disguise of police officers came upon them, and by the authority of forged orders and seals, confiscated their caravan, and burnt Iancu Dimonie alive on Petrina Mountain. Thus, Cova and Toli, Atanasi's children, were ruined, and remained the owners of a ramshackle inn on the road that went from Skopje to Ohrida.

Bankers, masons, tailors, shoemakers, silversmiths, fishermen, peasants, very good furriers and merchants. The Aromanians streamed in to Ohrida, and in order to make room for them, others left, thus taking away the burden of too many taxes.

They also say that the Aromanians of Ohrida had extensive building corporations, and in the summer time they went to work in Romania and Serbia. They returned home during winter. Gelagin Bey himself, the satrap whom Resid Pasha chased away in 1832, called on them and had them build a palace for him. So they *were* skilled. In 1836, they pondered a bit and came up with the idea of building a church. They chose Saint Gheorghe as their patron, and after a series of artful interventions for the necessary accreditation, Atanasi Dafin and another one walked for three years in search of sponsorship. And what they built during the day was crumbling during the night. They stayed on watch, and what did they see? The Turks kept messing with their walls! The Governor of Bitolia himself had to look into the business, and so the night guards were provided to make sure that the walls were ever to have a roof on top.

PERIVOLI

Some think that the most daring sort of Aromanians who lived in the Pind were the *caravanari* of Perivole. A caravan included 150 horses and mules, and the *caravanari* were bad omens to the Albanians. Costa Vraca had a very large caravan, but he was nothing like Gogu Musu. A patrol chief and *dervengiu*, the latter had become the aid of the Minister of Internal Affairs in Constantinople, and in 1860 participated in the Greek revolution of Crete. He was decorated upon three occasions and enjoyed many privileges. And here is note no. 7 on page 92 in Haciu's book, saying that Gogu Musu, together with other *celnici,* succeeded in unmasking a band of Turkish *caimacani,* who, disguised as Albanian thieves, looted the imperial treasury while on its way to Constantinople. The *caimacani* were captured and hurled into the Straits of Bosphorus.

What do you think Gogu Musu thought of in 1877? With help from the Metropolitan Patriarchy, he organized a revolutionary movement in the Pind. The Turks were upset, and from now on opinions diverge: according to some, he was exiled to Budrum in Minor Asia, while some others claim he was exiled to Berat. A third variant would have it that the British offered him protection and an important position in Cyprus.

PERLEPE

Several strata, one of Gramoste, another of Moscopole, a third of Crusova, and then who knows how many others... All together came to bear the name of Perlepe. This is the place where Theodor Capidan's father lived and sewed woolen coats. Theodor Capidan should be in schoolbooks, for every child to learn about.

SAMARINA

As was the case with Metovo, social strata also existed here: the *cogeabasi* and the plebeians who were called *cioplac*. The two classes despised one another and would confront on any occasion. During the 18th century, for instance, several families departed as a result of such confrontations. The defeated ones found refuge in the vicinity of Berat.

In Samarina, in the period between St. Gheorghe and St. Dumitru, a thousand weavers would work day and night. They had sixty felting machines and eighty mills. They also made packsaddles, girths, sandals, weapons, icons, bells, clothes and several other things, including Solingen knives. For distribution, they went to the fairs of Bitolia and Conita. When they went to the Conita fair, leading their burdened mules by the harness, children would come to meet them with icons in their hands, and the merchants would kiss them and give them money.

SARACU

Saracu was an important town, too, no less famous than Seatistea or Calaru. The noteworthy thing is that it used to be the capital of a confederation that, as early as the 17th century, counted 42 big villages. Its story ended as all of them did! Ali the satrap, the raving lunatic, came and swept away 75 families, that is almost 7,000 people. My opinion stays the same: had there been no Ali, they would have found some other occasion to leave.

Zalacostea the poet, whom the Greeks claim as one of their own, was born in Saracu. And so was Ali Pasha's doctor and counselor, Ion Coletti. Ion Ghica made the latter's acquaintance in Paris, in Mme Champy's salon. A Romanian and the ambassador of King Othon; and the ambassador said: *"I can speak Aromanian, but I am Greek."* Just like a shepherd of Brasov, Ghica commented. And the shepherd explained: *"My parents can only speak Aromanian, and I am glad you're one of us; we are sympatriots."* It happened around 1853-1856, and the two Romanian intellectuals who chanced to become acquainted in a Parisian salon began, predictably, to discuss books. Coletti asked whether there were any writers in the Principalities, and expressed his wish that he could read a book of Romanian poetry. Ghica sent him a copy of Iancu Vacarescu and Dionisise Fotino. It was from Coletti that Ghica found out about other Aromanians, those born into the Filiti and Cantili families.

Our Coletti had been brought up at Ali Pasha's court, by the side of some Christian children. With one of them, an Aromanian from Tricala, he shared the same room. One day, Ali was really furious because an important prisoner of his, Marco Botari the Suliot, had eloped. Assuming that the *tricaliot* was no stranger to the elopement, Ali started to yell at him, just as the young man was bringing in his morning cup of coffee. Whether he yelled too fiercely, or the boy was really guilty, one way or another, the hand trembled and the coffee was spilt all over the pasha's shalwars. Several minutes later the boy was decapitated. Frightened, Coletti decided to run away, but wasn't lucky and got caught. Meanwhile Ali's anger had died down. He was feeling sorry now, and lamented the boy's death – *crima to palicari*, he said – but it's all for the better that God took him, otherwise who knows what an evil-doer he would have turned into. As for Ali, that *was* a curious man: when Coletti was taken ill with typhus, he went into his room and, seeing how miserable he was, encouraged him and said, "you get well, boy, you just get well and Pasha will send you to Pisa where you'll become a doctor." And so he did. A tyrant all right, but kept his word when you least expected him to.

SARAJEVO

Sarajevo was the dwelling place of one Petre T. Petrovici (Petraki P.), who was a president of the church and school community for 20 years, then a vice-mayor and then, for another 20 years, the chief officer of the county. At the beginning of the 20th century, he had 4,000,000 *coroane* worth of fortune. In times of trouble he had given proof of a truly moderate and self-continent personality, and now the authorities appointed him as peacemaker whenever there was some popular upheaval. So, for instance, in 1878, when the Austrian-Hungarian armies came in, he did what he did and appeased Hagi Loji by offering him a fine, richly embroidered coat and a pair of silver guns. This is how he finally became a Star Commander of the Franz Joseph Order, received the 3rd class Iron Crown and the Golden Cross. He shared his mayor's salary out to the poor. Both he and his wife, born Economu, came from Moscopole. A descendent of this family became an Austrian consul to Adrianople, another to New York, and two others to less important cities, it doesn't matter where.

SKOPJE

A transplant of Crusova again. Haciu's book, on page 214, says: "When Constantin Casapana or Toma Dicea rode through this town, all the Turks rushed to stand up and give deep respectful bows."

Haciu also says, in his book: the cloth manufactory was like this, the tailoring was done like that; he also mentions, incidentally, one Cusu Haciu who, around 1900, was a tailor in Skopje. Nothing more on his own father who, as a young man, used to do business in Egypt, trading in wooden spools.

VELES

Among the families of Moscopole who were finally stranded in Veles was the Saru family. Lt. Boga told Haciu that his family had been the preserver of old Italian books and grammars and Greek manuscripts dating back to the 14th century. Professor Sava Saru showed Haciu a letter dated Feb. 2nd 1851, which makes it clear what Panait Saru's intentions were when he sent his son, Teodor, aged 11, to pursue his studies. Mr. Lica, the friend who was to be the boy's tutor, thanks Saru for his trust and says of the child, I quote: *"...he should be in good health, an honest and faithful young man, he should never lie, for he who lies is ready to give in to thievery too; he shall have to lend his ear to his father's and master's advice, to abstain from idleness, from straying here and there without my leave, he shall do whatever I will have him do, and I will help him improve, which will be to my greater joy. Therefore, if he is endowed with all these qualities, and if he has his mother's blessing too, and if he has a love of diligent study, you may send him soon after the Easter holiday and God bless us so we may succeed. Advise him to inform me on whatever he sees and hears, to be faithful to me and to show love towards my son."* Saru and Lica were therefore to live together as one family, and if Saru behaved and Lica had a marriageable daughter, when they came of age, it was customary for the tutor to give his daughter to his pupil to marry, so they could really become relatives after all.

VENICE

Valeriu Papahagi writes, and this is no surprise to me, that in Venice, the Aromanians succeeded in taking on high positions. They were, for instance, consuls and diplomats. He quotes Pietro Rose, who described the Aromanians as being *"nomini tutti de negocio e contegio capacissimi."* In 1719, G. Vretto was recommended to the Venetian ambassador to Constantinople as a diplomat of the consulate. He and Ioan Neranzi of Seacistea had been consular agents in Durazzo and Genova for England and, respectively, Holland.

Much as she might have treasured our people, I couldn't forget that, politically, Venice played a harlot's dirty game when she used the Aromanians to fight the Turks and then the Turks and *Uscoci* to beat down the Aromanian *armatoli* who encouraged French trading in the Balkans. This is how the Durazzo port was brought to rack and ruin.

But, in eating the other's heads, Venice actually ate its own head: starting with 1700, business started to decline, and even so, the *2 percent ad valorem* goods tax remained the same. I will not claim that Venice was an Aromanian town, I simply mean that Venice was yet another place where our presence was considerably felt. Under these circumstances, as Pietro Rose, the consul of the Venetian Republic I mentioned, understood how important the Aromanians were for the trade of the city, he spoke of them as I showed he did, and did whatever was in his power to make them give up the shipping with Dulcigho ships – which was the loathed enemy of the lagoon city. But what reason did the Aromanians have to side with a potential loser?

VERIA

Now hear this one, it'll be about Barda again. At the beginning of the 19th century he brought a group of Aromanians from Avdela and chose a dwelling place for them somewhere around Veria, not far from the rock where Apostle Paul preached, so that a lot of boys in the whereabouts came to be called Paul.

Our neighbors in Doliani were called Caranica, Badralexi, Caprini, Tanasoca, Tosca, Dauti, Stamboli, Busulenga, Bucovala, Samara, Vrana, and Pitulia… My dear old grandma found solace in that, although a widowed and poor woman, she was born into the fine line of the Toscas and had become in-laws with a Samara. Not far from us was Gramaticova, where Tal Celea's descendants came to live. Gusu Celea, one of them, together with Tea Cusa moved heaven and earth to raise a newer and handsomer Gramaticova, which people started to leave after twenty years.

ZAGORE

A cluster of small towns and villages, around forty of them, spread all over the Zagor dip. Around 1930, only some thirteen of them were left untouched by the "Greek turn". It had once been the estate of Salik Bey, the third son of Ali. The Aromanians of Zagor were in big business with Egypt, Minor Asia and the Romanian Principalities, where they were into agriculture – can you imagine that? Just as, whenever a boy was born into a Phanariot family in Constantinople, the wishing was: *may he grow up a fine man and become a prince in Moldavia or Vallachia*, in Zagore, on similar occasions, they said: *may you grow up, make your fortune and become a pastry cook in Romania*.

In many of their houses you could find Mihai the Brave's portrait hanging on the walls. In 1984, I could record the song of Mihai Bey, which the Aromanians sang in Greek and T. Burada published in *Convorbiri literare* ["Literary Conversations"], as sung by *lala* Chichi and Halciu al Mani, and I still listen to it from time to time.

A strange sort of "specialization" divided these 40 villages, here are some examples: Leasnita gave priests and teachers, Cernes church painters, Gubenite merchants, Floru photographers and pastry cooks, Leascova *caloiatri* medical doctors. On 16th-century Turkish maps the Zagora area was called Olah Kioi. Monk Cosma tore into Olah Kioi like a whirlwind!

EPILOGUE

Not far from us, here, in the Balkan Peninsula, a millenium of history has been missed out! It's been a waste which the waster himself indulged when he failed to allow the act of confession its due place within the range of his manifestations. He who omits speaking about himself will only be remembered in history from things others care to put on paper.

When psychoanalysis has demonstrated that not only people, but cultures, too, suffer from complexes, the diagnosis I propose now will no longer be a simple metaphor. I'll say that the Aromanian culture suffers from the Atlantis complex! This complex is typical of those cultures that, in responding to drives unknown to us, are present in written history by their absence.

The picture I can form of the Romanian nation as I look at it out of this personal window of mine is that of an egg with two yolks, in between which the Danube flows. And the Aromanians appear to me as a paradoxical ethnic race, which shone bright in the night of the Middle Ages, but for the last two centuries has been slowly dying out, as it has gradually entered the cone of light of the Present.

An ethnic race dies at its own cost. It leaves behind objects, songs, dances, beliefs, stories that become useless, sometimes even ridiculous. The death of an ethnic race occurs when the people choose another culture. Had the Aromanians abandoned their culture, they would have acted like Ungaretti's Ulysses, who never returned to Ithaca. But their lot, as they were always traveling and wandering about, was to come back and no longer find Ithaca, or in other words, their world dissolved before they had even left it. So they gathered in one place whatever each of them had on himself. Nowhere is it written that the Phoenix looks the same after it is born again, rising from its ashes.

...having a mind to write about the Aromanians, I intently dreamt of a book that should be consubstantial with their world. I've built it piece by piece, I've spurred it, taken it to pasture, combed and polished it. It was meant to be created after their likeness, from the waist upwards it was wool and salt and from the waist downwards it was mountain paths. The pages piled up as on a bazaar stall, vying for supremacy by invoking words with archetypal undertones: road, sheep, gold, freedom, cheese, gun powder...

This is the end, I've said it all, as well as I could. Whether it's much or little, it all depends on how you read and what you expect. There's been exaggeration and there's been simplification, I've been passionate but I haven't kept things from you, every word is a picture of myself. Whoever resented my story will kindly forgive me, but those of you who were pleased with it, here's the cap, leave a coin for the teller!

GLOSSARY
by Sorana Corneanu

This is a very brief introduction to some of the Romanian words in the text which may be partially or totally unfamiliar to the foreign reader. We have left some others unexplained here, since the text itself makes them intelligible either at the place of their first occurrence or in some other section. A minimal measure against dry formality was to give this quasi-random listing of the most important terms (grouped according to some more or less loosely unifying features) – a fragmentary guide to what the reader is invited to imagine as yet another book of stories.

A *caimacam* was the deputy of a prince.

The *Dieta* was the legislative assembly of the state.

A *dervengiu* was someone in charge with the safety of some mountain pass (*dervena*).

A *zapciu* was a watchman.

Eteria was a Greek patriotic society formed with a view to organizing anti-Ottoman resistance.

An *exarhat* was a part of the Roman Eastern Empire under the ruling authority of an *exarh* – a metropolitan bishop deputed by the Patriarch of Constantinople to rule over the church of a certain province or country.

The *Bairam* is the period of celebrating time preceding or concluding the Mohammedan Ramadam. The *Ramadam* is the most important fasting period with the Moslems, which extends over the ninth month of the year.

A *bey* was the governor of a Turkish province or town (less important then a pasha) and also the title given by the Turks to Romanian princes.

A *cadi* is a Moslem judge.

A *hoge* is a Moslem priest, and so is an *imam*.

A *hamam* is a dry-air Turkish bath.

A *giamie* is an Islamic temple.

A *valiu* was the governor of a small province in the Ottoman Empire.

A *bajbuzuc* was an Albanian soldier.

A *bogomil* is an adept of *bogomilism* – a medieval religious doctrine and sect established in Bulgaria, which acknowledged neither the precepts of the Christian church nor the authority of the state.

A *rudar* is a Gypsy woodworker.

If someone was said to be a *cogeabas*, it meant he was an important and respectable person. The name of *everghet(a)* was given to someone who generously offered important sums of money for a good cause. A *hagiu* was someone who was a pilgrim to holy places (in Jerusalem or Mecca); it was also the title that person received when he returned from his journey.

Hora is a Romanian traditional dance where the dancers hold one another's hands and form a full circle.

Farserot woman

Men and woman of Smixi

Photographer Ionel D. Manachia of Bitolia

Woman of Vlaho-Zohar

Old costume of Avdela

Woman of Laca-Zagor

Type of Perivole

Man of Metova

Hero Mihali Tegu Iani of Perivole

Mayor M. Dida of Magarova

Old man of Smixi-Pind

Albanians of Dibra

Bride and groom of Samarina

Bride and groom of Blata

Bride and groom of Turia

Type of Pazi

Meglenit woman of Liumnita

Woman of Abela

Party at the spring, Avdela

Wedding party in Smixi

Hora dance in Smixi

Heroes of Perivole

Caravan men of Pind

Man of Perivole

The Tegu Iani family of Perivole

Farserot women

The George Pupi family of Avdela

Types of Calari

Farserot types

Baeasa commune

The Romanians School of Ianina

Aromanians on the road

Magarova commune

Type of Pazi

Crusova commune